Modeling Our World

SECOND EDITION

The ESRI Guide to Geodatabase Concepts

MICHAEL ZEILER

ESRI Press
REDLANDS, CALIFORNIA

Cover image from ESRI Data & Maps 2008, courtesy of NASA, Tele Atlas North America, and ArcWorld Supplement

ESRI Press, 380 New York Street, Redlands, California 92373-8100

Copyright © 1999, 2010 ESRI

All rights reserved. First edition 1999. Second edition 2010.

14 13 12 11 10   1 2 3 4 5 6 7 8 9 10

Printed in the United States of America

Library of Congress Cataloging-in-Publication Data

Zeiler, Michael.
    Modeling our world : the ESRI guide to geodatabase concepts.—2nd ed.
    p. cm.
    ISBN 978-1-58948-278-4 (pbk. : alk. paper) 1. Geographic information systems. 2. ARC/INFO.
3. Digital mapping. I. Environmental Systems Research Institute (Redlands, Calif.) II. Title.
    G70.212.Z45 2010
    910.285'574 dc22   2010017004

Ask for ESRI Press titles at your local bookstore or order by calling 800-447-9778, or shop online at www.esri.com/esripress. Outside the United States, contact your local ESRI distributor or shop online at www.eurospanbookstore.com/ESRI.

ESRI Press titles are distributed to the trade by the following:

In North America:
Ingram Publisher Services
Toll-free telephone: 800-648-3104
Toll-free fax: 800-838-1149
E-mail: customerservice@ingrampublisherservices.com

In the United Kingdom, Europe, Middle East and Africa, Asia, and Australia:
Eurospan Group
3 Henrietta Street
London WC2E 8LU
United Kingdom
Telephone: 44(0) 1767 604972
Fax: 44(0) 1767 601640
E-mail: eurospan@turpin-distribution.com

Contents

Foreword

In 1999, Michael Zeiler released his first edition of *Modeling Our World*. This coincided with the initial release of ArcGIS and the geodatabase. His book was an excellent primer for many to begin to understand the fundamentals of GIS and how geographic information was represented and used in a GIS. His work focused on the fundamental geographic information concepts underlying the geodatabase and its use.

Those were heady days. The use of relational databases and large multiuser enterprise systems, while still a relatively new approach, was growing rapidly. The GIS community also was changing dynamically, experimenting with all kinds of approaches for representing and managing geographic information. As a result, many people needed real help and knowledge to make their new GIS implementations work.

The first edition of *Modeling Our World* filled a huge void. For the first time, many new users began to understand fundamental information concepts about GIS. Michael's book excelled in its use of graphics to illustrate, simplify, and communicate an array of complicated subject matters. It brought to life many of the essential GIS data concepts. His graphical approach provided important insights about geographic data modeling. His book made a big imprint on the successful use of GIS, lifting the skills and knowledge of the rapidly expanding GIS community. It helped academics begin to teach GIS, which in turn enabled legions of new users.

Michael's career has focused on communicating complex ideas through graphical representations. His goal has been to create pictures that enable people's ability to understand and grasp complex subjects and concepts. His mission has been to enlighten and teach by combining explanations and concepts with beautiful diagrams and examples that engage readers and spur their imaginations.

As GIS continued to expand during the last decade, we began to recognize—and to be humbled by—the progress that was happening in hundreds of thousands of organizations worldwide. Collectively, the GIS community was developing incredible examples and knowledge about geographic information representation and modeling.

At ESRI, we sensed that the time had come to write and share a new edition of *Modeling Our World* for the next generation of GIS professionals. We wanted to incorporate the many lessons we had learned, along with the many examples of GIS implementation. We started from first principles. We put a strong team together from the geodatabase group at ESRI and invested incredible amounts of energy to capture and articulate new insights about geographic information and how geodatabases are used.

For this new edition, Craig Gillgrass, Jonathan Murphy, Matt McGrath, and Brent Pierce teamed with Michael to focus on clarifying and communicating this important body of information and the lessons learned. The second edition of *Modeling Our World* is about enabling GIS practitioners to implement effective, elegant systems. People and their collective knowledge are a fundamental component in every geographic information system. Professional GIS users have the role of creating and managing authoritative geographic information for their subject matters and their areas of interest. We think that *Modeling Our World* will enable these professionals to do their jobs better.

It is inevitable that these information collections will be accessed and combined on the Web in ways that none of us can yet imagine. At ESRI, we strive to provide a framework that enables users of all types and levels to participate in a shared GIS. We all sense that GIS will have an enormous impact on our lives and our abilities to manage our planet. Our strong hope is that this book will enable you to successfully use and apply GIS in your many and varied endeavors.

Clint Brown, Director of Software Products
ESRI
Redlands, California
May 2010

Acknowledgments

This book, *Modeling Our World,* is the collaborative effort of many people's inspirations, ideas, and labor.

Many ESRI employees deserve recognition for their groundbreaking work on bringing ArcGIS release 10 to the ESRI user community; this book documents the fruit of their expert knowledge, creativity, and long hours of effort. Many persons within the ESRI user community also contributed to this book through their excellent maps and applications. Because of space constraints, I can only acknowledge those persons who most directly participated in the creation of this book.

A group within ESRI's geodatabase team worked very closely with me to make this book relevant, accurate, and useful to the GIS professionals who pick up this book. Jonathan Murphy led this group and made extensive edits and contributions throughout the book. Brent Pierce lent his expertise about modeling features in the geodatabase and was a key contributor to the chapter on versioning. Craig Gillgrass applied his broad knowledge of the geodatabase and was a key contributor to the content on geometric networks. Matt McGrath coordinated this review team and applied his organizational talents.

Many product engineers and programmers at ESRI provided content for this book and reviewed chapters. Melita Kennedy and David Raleigh are experts in coordinate systems and spatial references and helped me with chapter 2. Derek Law provided content for understanding coordinate precision and resolution. Charlie Frye and Aileen Buckley work in ESRI's cartography lab and helped with map publication concepts for chapter 3 and elsewhere in the book. Alan Hatakeyama, Matt Crowder, Craig Gillgrass, and Robert Garrity provided their network modeling expertise for chapter 4. Heather McCracken lent her expertise on linear referencing and Adrien Litton, Ray Carnes, Eric Floss, and Terry Bills did chapter reviews. Agatha Wong provided her long experience with geocoding, with additional content and reviews by Jeff Rogers, Bruce Harold, and Lucy Guerra. Melanie Harlow is an expert on rasters and imagery and provided much of the content of chapter 7. Peter Becker and Lawrie Jordan offered their expert knowledge in bringing chapter 7 up-to-date with the latest imagery technology and Hong Xu provided a chapter review. Clayton Crawford and Lindsay Weitz applied their deep knowledge of terrains and surface modeling to chapter 8. Hardeep Bajwa, David Kaiser, Morakot Pilouk, and Steve Kopp gave concepts and details for temporal modeling concepts in chapter 9. Brent Pierce and Jonathan Murphy were the key contributors for versioning and replication concepts in chapter 10. David Wynne, Scott Murray, Dale Honeycutt, and Steve Kopp applied their knowledge of geoprocessing models and scripts for chapter 11. Erik Hoel is the geodatabase software architect and he answered many questions.

Several GIS experts in the ESRI user community provided invaluable content. The City of Santa Fe and County of Santa Fe graciously offered their GIS data that is used extensively for map illustrations in many chapters of this book. I worked with Denise Vigil and Leonard Padilla at the City and Amanda Hargis and Erle Wright at the County. Ken Logsdon of Dewberry Inc. gave maps and best practices for modeling terrains from lidar data. At Public Service of New Mexico (PNM), Ted Kirchner, Gathen Garcia, and Chris Carpenter provided maps and their insights on how versioning and replication is applied at an electric utility. And I would like to thank the many persons within the ESRI community who made the beautiful maps that appear throughout the book.

Several persons participated in the production of this book. I consulted with the team at ESRI Press on the book design and permissions work, and they applied their collective experience to preparing this book for publication. Joyce Frye did the copyediting of the book and corrected many errors; any that remain are mine.

The production of this book is due to the vision of ESRI's management. Clint Brown is the Director of Software Products and took a special interest in ensuring that this book serves the conceptual requirements of the ESRI user community. Scott Morehouse is the Director of Software Development and is ESRI's visionary on advancing the theory and practice of GIS. Peter Adams leads ESRI Press and directed the publication of this book. Jack Dangermond created this very special and unique institute where we can believe that we make a difference in this world and act on that idea.

Finally, I give very special thanks to my wife, Polly White, for her endless love, support, and encouragement. She and my children, Petra and Maximilian, displayed much grace and patience during the long hours that that it took to make this book.

- Michael Zeiler

Modeling Our World

SECOND EDITION

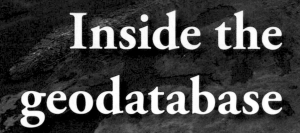

Inside the geodatabase

A map portrays geographic information through a series of thematic overlays—streets, buildings, water, terrain, imagery. So does a geographic information system (GIS). Each geographic data model is a collection of thematic overlays used to represent geographic location and shape as well as descriptions of the geographic features (points, lines, polygons, imagery, and so on). The geodatabase is a collection of these datasets and properties.

The geodatabase extends tables in a relational database with shape fields that contain point, line, or polygon geometries. Imagery and continuous data are stored with efficient raster data structures scalable to large data volumes. From these simple tables, feature classes, and raster datasets, many structures and techniques—feature datasets, subtypes, attribute domains, networks, terrains, address locators, and more—provide a framework for sophisticated modeling of natural and man-made systems.

A map is how people share geographic information. Take a close look at this remarkable map from 19th century Japan designed to guide visitors through Osaka.

This map uses several cartographic techniques for thematically representing the city:

- City blocks are drawn with colors coded by ward.

- Waterways are mapped with parallel lines and symbols of ships.

- Place names are annotated throughout the map.

- Feature types are classified through a map legend.

- Addresses and distances are shown in tables referenced to the map.

Many layers of rich information combine to produce this definitive city map. By reading this map, a visitor can locate any place in Osaka by address, and find the best route and distance to that place.

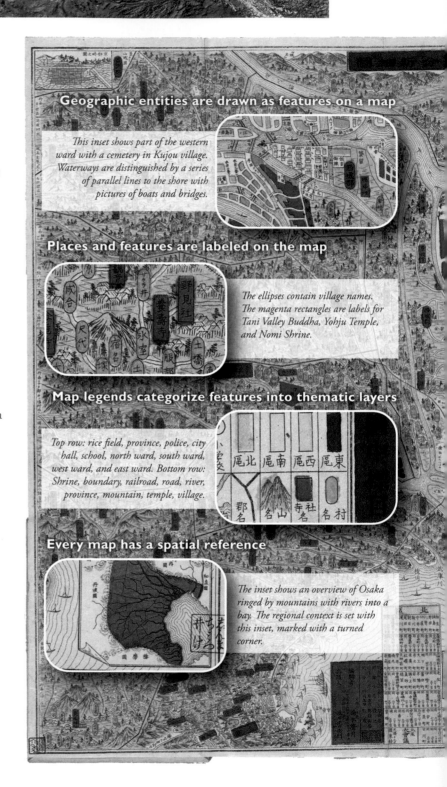

Geographic entities are drawn as features on a map

This inset shows part of the western ward with a cemetery in Kujou village. Waterways are distinguished by a series of parallel lines to the shore with pictures of boats and bridges.

Places and features are labeled on the map

The ellipses contain village names. The magenta rectangles are labels for Tani Valley Buddha, Yohju Temple, and Nomi Shrine.

Map legends categorize features into thematic layers

Top row: rice field, province, police, city hall, school, north ward, south ward, west ward, and east ward. Bottom row: Shrine, boundary, railroad, road, river, province, mountain, temple, village.

Every map has a spatial reference

The inset shows an overview of Osaka ringed by mountains with rivers into a bay. The regional context is set with this inset, marked with a turned corner.

The intention of this (and every map) is to transmit geographic knowledge. Maps use a variety of graphical techniques employing size, color, symbol, and annotation to display information about what is at every place within the map. The purpose of a map is to show information useful to the map reader's tasks.

Maps are designed to support spatial analysis in the mind of the map user. Today's geographic information systems also produce maps to support the map reader's goals, but with the advantage of computer technology.

With powerful GIS displays and analytical functions, you can think of a modern geographic information system as a telescope for geography.

Courtesy of the C. V. Starr East Asian Library, University of California, Berkeley

Think of a map as a geographic information system that you can fold and put in your pocket.

Maps work because they follow a well-evolved set of practices to present geographic information.

Maps present a view of a geographic area. Representations of real-world entities are drawn with point, line, and polygon symbols and are also shown with imagery.

Points, lines, polygons, and images on a map are organized into thematic layers. A collection of points can be used to represent cities or well locations. Lines show features such as road systems or river networks. Polygons describe areas and are often used to depict buildings, states, or countries. Imagery shows a picture of the landscape captured from airplanes or satellites.

Maps apply symbols thematically. Rivers are typically blue lines, roads may be red or black lines. City blocks might be color coded by zone. The map legend is a symbolic inventory of all the types of features in a map.

Thematic layers are the starting point for modeling our world. Maps combine many thematic layers over a common geographic area. These thematic layers not only contain collections of similar entities, but also model their attributes, relationships, and behavior.

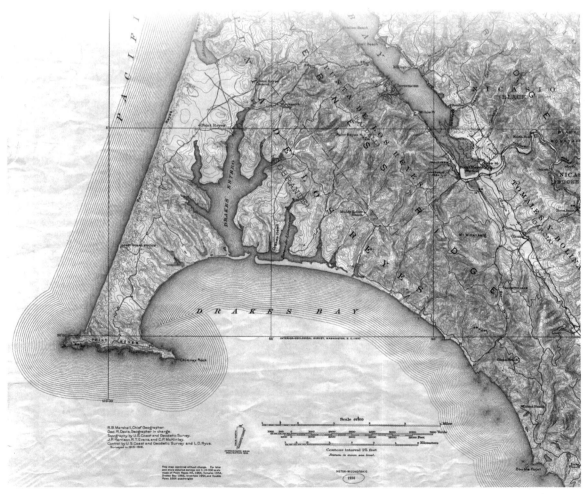

This USGS 7.5 minute quadrangle map of Point Reyes, California from 1916 demonstrates the use of point, line, and polygon symbols.

The same elements and principles used to create maps are employed by ArcGIS to construct modern geographic information systems.

In ArcGIS, the geodatabase is the native data structure for the storage and analysis of geographic information. Just like maps contain a collection of many thematic layers, the geodatabase is a collection of thematic datasets.

Each dataset stored in a geodatabase can be thought of as a thematic layer which is used to represent real-world entities. Just like the thematic layers on a map, each dataset in a GIS is a layer —a logical collection of features such as wells, parcels, rivers, roads, and so on. Like the map projection information defined for a map, each dataset in a geodatabase has a spatial reference. This spatial reference is used to position thematic layers when they are overlaid on a map.

Datasets representing satellite and aerial photo imagery are also stored in a geodatabase. These are used as a foundation for other thematic layers or to represent continuous surfaces and thematic data. Imagery is georeferenced so that it is also aligned with other datasets.

Geographic information systems reference geographic datasets stored in a geodatabase by overlaying these datasets as thematic layers on a map. In this sense the geodatabase is a key aspect to modern geographic information systems.

Geodatabase
Mexico

Point feature class
City

The cities of Mexico are drawn by population with circle symbols graduated both in size and color.

Line feature class
River

The major rivers of Mexico are drawn with lines color coded for river classification.

Line feature class
Highway

The highway system of Mexico is drawn with colored lines for highway types.

Polygon feature class
State

The United Mexican States (Estados Unidos Mexicanos) are shaded with random colors and labeled with their name.

Raster dataset
EarthImagery

The land is captured in this raster mosaic combining many satellite images to produce a cloud-free view of the landscape.

In its simplest terms, the geodatabase is a collection of geographic datasets stored using a database management system (DBMS) or file system. However, the geodatabase has several other key aspects:

- The geodatabase is the native data structure for ArcGIS and the primary data format used for editing, representing and managing geographic data in ArcGIS.

- The geodatabase has a data model that is implemented as a series of simple data tables holding feature classes, raster datasets, and attributes. In addition, advanced GIS data objects add GIS behavior, rules for managing spatial integrity, and tools for working with numerous spatial relationships of the core features, rasters, and attributes.

- The geodatabase has a transaction model for managing GIS editing workflows such as adding, deleting, and updating features and attributes.

- The geodatabase provides a common foundation for accessing and working with all geographic data in a variety of files and formats.

The geodatabase is designed to represent real-world entities using imagery and simple features: points, lines, and polygons. Various types of geographic datasets can be created using these simple features. There are three fundamental geographic datasets used in a geodatabase: tables, feature classes, and raster datasets.

- Tables are used to manage descriptive information as attributes, such as records of ownership, measurements at a location, or qualities of objects.

- Feature classes are thematic datasets which represent geographic features such as parcels, utility lines, and wells. Feature classes are tables with a spatial field. An additional column with a shape field is used to specify a geometry of point, line, or polygon.

- Raster datasets represent imagery of the earth and other continuous surfaces. Each raster image is stored as a distinct thematic layer.

It is from these fundamental geographic datasets that you will begin to build a GIS. As the requirements for your system extend past this fundamental information model, the geodatabase can adapt through application logic used to extend the capabilities of these datasets to model spatial relationships, improve data integrity, and add dynamic behavior.

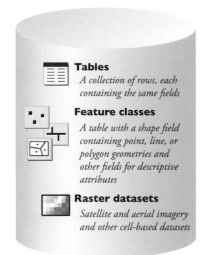

Tables
A collection of rows, each containing the same fields

Feature classes
A table with a shape field containing point, line, or polygon geometries and other fields for descriptive attributes

Raster datasets
Satellite and aerial imagery and other cell-based datasets

There are three types of geodatabases: personal, file, and ArcSDE. Selecting the appropriate geodatabase to work with will depend on the specific requirements of your GIS project.

Personal geodatabases

The personal geodatabase is based on Microsoft® Access® software and was the original geodatabase available at the launch of ArcGIS 8. It is designed for single users working with small to moderate size GIS datasets. The personal geodatabase is implemented within a single Microsoft Access file with a maximum size capacity of 2 gigabytes.

File geodatabases

The file geodatabase is ideal for GIS projects done by single users and small workgroups. The file geodatabase improves on the personal geodatabase in terms of scalability, performance, and cross-platform operating system support. File geodatabases scale up to handle very large datasets by using an efficient data structure that is optimized for performance and storage.

The file geodatabase was introduced at ArcGIS 9.2. The file geodatabase is implemented as a collection of binary files in a file system and because of this has no practical size capacity limit. Each table in a file geodatabase can be configured to store up to 256 terabytes of information.

Vector data in file geodatabases can be compressed to a read-only format, providing a smaller memory footprint and improved geodatabase performance. Uncompressing the vector data will make it editable.

The file geodatabase can support more than one editor at a time, provided the editors work on different tables, stand-alone feature classes, or feature datasets.

For new geodatabases, ESRI generally recommends file geodatabases over personal geodatabases because of improved functionality and better performance.

ArcSDE geodatabases

ArcSDE geodatabases are designed to be scalable and to support multiple readers and writers concurrently. Each has the ability to manage a shared, multiuser geodatabase as well as support for a number of critical version-based GIS workflows such as access permission control for individual datasets, versioning, geodatabase replication, and historical archiving. ArcSDE geodatabases can be implemented on many different relational database platforms including Microsoft SQL Server®, Oracle®, IBM® DB2®, IBM Informix® and PostgreSQL. Because the enterprise level of geodatabases use relational databases, they are only limited by the underlying relational databases and can be scaled to any size and support any number of users, running on computers of any size and configuration. The ability to leverage your organization's enterprise relational databases is a key advantage of the ArcSDE geodatabase.

The geodatabase has several key qualities that make it flexible enough to manage any GIS.

1 Simplicity and efficiency

The geodatabase represents real-world entities in a simple and efficient way. The geodatabase is built on a base *relational model* in which geographic data is organized into tables which contain rows. All rows in a table have the same columns, therefore maintaining relational integrity. Each column has a type such as integer, decimal, character, date, and so on.

The geodatabase extends this relational model by including a column with a geometry type. This allows the geodatabase to conform to a simple feature model where geographic elements in the real world are represented by simple features with geometries of point, line, and polygon. Using this simple and efficient design allows the geodatabase to be fast and scalable.

2 Transactional model for editing workflows

The geodatabase also has a comprehensive *transactional model* used to apply edits to rows and columns in tables. This allows users to perform inserts, updates, and deletes on data. The geodatabase can extend this transactional model to accommodate a wide range of workflows including multiuser editing and data distribution through geodatabase replication.

3 Scalable framework

The geodatabase is a robust and customizable framework that can scale from small, single-user databases up to large enterprise-level systems with many concurrent editors. Because of the simple design behind extending the fundamental relational, simple feature, and transactional models, the geodatabase has the flexibility to represent geographic data models of any size or design.

4 Effective display and analysis of geographic data

Datasets stored within the geodatabase can be overlaid on a map for display and analysis. Combining thematic datasets over a common geographic area helps you visualize how the layers interact. Datasets can be used as input into geoprocessing tools and models to perform analyses. Results can be displayed on the map or even used to derive new datasets which can be used for further analysis.

5 Extended capabilities and behavior

Simple features stored in tables are the foundation of the geodatabase's information model. Groups of points, lines, or polygons are combined into feature classes. Tables, feature classes, and raster datasets form the basis for representing real-world entities as geographic data. The geodatabase can extend this information model with more advanced dataset types, capabilities, and application logic used to maintain data integrity and model spatial relationships. The geodatabase can also model more dynamic systems such as topologies, networks, parcel fabrics, and surfaces.

Since the advent of computer-based geographic information systems in the 1960s, many different formats for storing geographic data have been developed. These geographic data formats include coverages and shapefiles, which were developed by ESRI for previous generations of GIS software. While coverages and shapefiles were very successful in their time and to this day have widespread use, they now place technical limits on users for modeling geographic systems and handling huge amounts of data.

In the year 2000, ESRI introduced the geodatabase with the release of ArcGIS 8. The geodatabase was created in response to user requirements for increased scalability for large datasets, open access to data, and richer modeling of geographic systems such as transportation networks, surfaces, and parcel cadastres.

The creation of the geodatabase became possible at that time because relational database technology had matured to the point that extensions to relational database technology could handle the specialized requirements of geographic data. Since then, subsequent releases of ArcGIS software have continually added enhancements to the geodatabase and it remains the state of the art.

6 Flexible storage and open access

Geodatabases can be stored as files in most relational database management systems (RDBMSs) and can be openly shared using Extensible Markup Language (XML). Because of their simple storage model, it is very easy to transport spatial data to and from geodatabases.

7 Data accessibility

The geodatabase provides a number of ways to access geographic datasets. These range from using ArcGIS applications such as ArcCatalog and ArcMap to programmatic access using one of the many geodatabase application programming interfaces (API).

To promote accessibility and integration with different GIS implementations, geographic datasets are also accessible through database-level Structured Query Language (SQL).

8 Interoperable with data from many sources

The simple feature model also allows the geodatabase to be interoperable with many different data sources that are not native to ArcGIS. The geodatabase can import and export geographic data from all significant GIS formats because other geographic data sources adhere to a similar simple feature model. There are tools and processes designed to migrate any existing geographic data into the geodatabase.

The geodatabase stores information using relational database principles. Whether implemented within an instance of a relational database or as files within the file system, the geodatabase has a simple and consistent information model for storing and accessing information in tables.

The fundamental idea is that data is organized into tables, which consist of a set of rows representing similar objects. Each row in a table shares a common set of fields (or columns) and each field has a type such as integer or text.

Storing attributes

All three primary datasets in the geodatabase (feature classes, tables, and raster datasets) as well as other geodatabase elements are stored using tables in a database. These tables are related to one another using key values in the attributes of the tables. This is commonly referred to as the relational model and is the backbone of the relational databases on which the geodatabase is built.

Tables also have integrity rules which help ensure the correctness of data stored in a table. Other types of rules control how objects can be related to each other, either logically or spatially.

Storing feature shapes

A geodatabase stores geographic objects like any other attribute. The geodatabase adds geographic data to relational database tables by adding a column to a table which stores the shape of geographic features. This allows the geodatabase to assign a value of point, line, or polygon to a feature. This column is the geometry type of a feature and is called the shape field in a feature class. In this way, a feature class can be thought of as a standard table with a spatial column.

Object-relational databases

The geodatabase is often thought of as an object-relational database because it employs a multitier application architecture. The advanced logic and behavior is implemented in the application tier which is separated from the data storage tier controlled by the relational database. The geodatabase application logic includes support for a series of generic GIS data objects and behaviors such as feature classes, raster datasets, topologies, networks, address locators, and much more.

Responsibility for management of geographic datasets is shared between the ArcGIS software and relational database. Certain aspects of geographic dataset management, such as disk-based storage, definition of attribute types and associative query processing, are delegated to the data storage tier in the relational database. The GIS application tier retains responsibility for defining the specific schema used to represent various geographic datasets and for domain-specific logic, which maintains the integrity and utility of the underlying records.

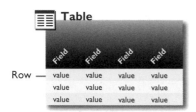

Table

Row

A table is the primary unit of storage in a database. Conceptually, a table is a matrix of values organized into rows and columns. Each row represents an entity. Each column or field represents an attribute type shared by all rows.

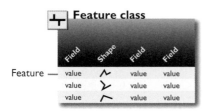

Feature class

Feature

A feature class is simply a table with a special field for storing geometric shapes of features. The three main types of features are point, line, and polygon.

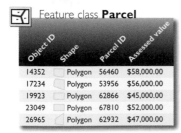

Feature class **Parcel**

Object ID	Shape	Parcel ID	Assessed value
14352	Polygon	56460	$58,000.00
17234	Polygon	53956	$56,000.00
19923	Polygon	62866	$45,000.00
23049	Polygon	67810	$52,000.00
26965	Polygon	62932	$47,000.00

Each row in this parcel feature class has a shape field for the parcel boundary and the attributes of that parcel.

This could be considered a multitier architecture (application and storage), where aspects related to data storage and retrieval are implemented in the data storage tier as simple tables, while high-level data integrity and information processing functions are retained in the application tier (ArcGIS). All ArcGIS applications interact with the application tier and the GIS object model for geodatabases, not with the actual SQL-based relational database instance.

Information in the geodatabase is stored in two tiers. In the storage tier, data is stored in simple tables managed by a database management system. In the application tier, data is accessed through ArcGIS applications as tables, feature classes, and rasters.

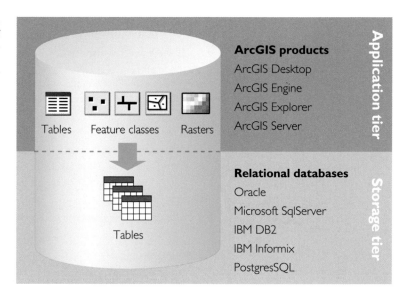

The three fundamental datasets in the geodatabase—tables, feature classes, and raster datasets—are the primary mechanism used to organize and use geographic information in ArcGIS. Using these primary datasets to model your GIS data is only the first step in building a geodatabase. Various geodatabase elements can be used to extend tables, feature classes, and rasters to model spatial relationships, add rich behavior, improve data integrity, and extend the geodatabase's capabilities for data management.

Extending tables

Tables provide descriptive information for features, rasters, and traditional attribute tables in the geodatabase. Users perform many traditional tabular and relational operations using tables.

In the geodatabase, there is a focused set of capabilities that are optionally used to extend the capabilities of tables. These include the following:

Attribute domains allow you to specify a list of valid values or a range of valid values for attribute columns. Domains should be used to help ensure the integrity of attribute values such as enforcing data classifications (road class, zoning codes, and land-use classifications).

Relationship classes build relationships between two tables using a common key and allow you to find the related rows in a second table based on rows selected in the original.

Subtypes manage a set of attribute subclasses in a single table. For example, a road feature class may have subtypes for divided highways, major roads, minor roads, and so on.

Extending feature classes

Each feature class in a geodatabase is a homogeneous collection of geographic features with the same geometry type (point, line, or polygon), attributes and spatial reference. Feature classes can be extended in the same manner as tables, but can also be extended through additional methods to support more advanced GIS workflows:

Feature datasets hold a collection of spatially related datasets that share a common coordinate system and are used for building topologies, networks, terrains, or parcel fabrics.

Topologies model rules for how features share geometry. For example, a set of land parcels should have polygon shapes that combine to fill an area without gaps or overlaps.

Network datasets are used to model transportation connectivity and flow.

Geometric networks are used for the editing and tracing of networks used for facilities management.

Terrains represent the surface of the earth with triangulated irregular networks (TINs) and are built from point, line, and polygon features that contain elevation values.

Annotation names and describes features with text on a map.

 Dimensions are a special kind of map annotation for showing lengths or distances on a map.

 Linear referencing locates events along linear features with measurements such as milepost data for roads or borehole depth.

 Multipatch features represent the three-dimensional shape of features such as buildings or subsurface geological units.

 Parcel fabrics model the spatial relationships for lot lines, parcel corners, and parcels as an interconnected fabric. They hold survey information for subdivisions and parcel plans.

 Schematic datasets allow you to visualize network, geographic, and schematic representations in the same environment through a graphic representation of your network connectivity.

Address locators specify how address information is organized and the street data from which places are interpolated.

Cartographic representations manage multiple symbolizations of features using rules for advanced cartographic drawing and placement.

Extending raster datasets

The geodatabase can manage rasters for many purposes: as individual datasets, as logical collections of datasets, and as picture attributes in tables.

Rasters are heavily and increasingly used in GIS applications. Most often, raster datasets store photographic maps from aerial photogrammetry or satellite imagery. Another use for raster datasets are gridded data such as digital elevation models. Raster datasets can also store scanned map data, documents, and photographs of features. A number of geodatabase capabilities enable users to extend how they manage their raster information as follows:

 Raster datasets are used to manage very large, continuous image datasets, such as a mosaicked image of a metropolitan area.

 Raster catalogs are used to manage imagery for a number of purposes including a tiled image layer, where each tile is a separate image, and any series of images in a DBMS or a raster time series.

Raster attributes are used to embed photographs or scanned documents as attributes within feature classes and tables.

 Mosaic datasets are related collections of raster and image data. Each represents an on-the-fly mosaic view of a raster catalog. Mosaics have advanced raster querying capabilities and processing functions and can also be used as a source for image services.

THE GEODATABASE TRANSACTION MODEL

GIS users have some specialized transactional requirements. Often, transactions must span long periods of time (sometimes days and months, not just seconds or minutes).

In a multiuser database, the GIS transactions must be managed on the relational database's short transaction framework. ArcSDE technology plays a key role during these operations by managing the high-level, complex GIS transactions on the simple transactional framework of a relational database.

Geographic data is used to model many types of real-world entities. Although the characteristics of these entities can vary, they have one quality in common—they are dynamic.

Consider a land parcel feature class as an example. Over time, land parcels undergo changes such as splitting or merging boundaries, reassignment of zoning codes, and transfers of landownership. Because geographic information changes over time, the geodatabase provides a transactional framework for users to easily edit their datasets while preserving data integrity.

When you edit geographic data, you follow a workflow similar to editing text within a Microsoft Word document. As you write a word document, you add, update, and remove text and pictures. At any time, these modifications can be undone, which restores the Word document to a previous state. The edits you perform are only committed when the Word document is saved.

Likewise, when you edit in ArcGIS, you add, update, and delete features in your geographic data. Changes to datasets in your geodatabase are committed only when you save your edits.

ArcGIS extends this simple transactional model to accommodate more advanced GIS workflows as well. Organizations such as utilities have complex transactions that span days or weeks. These are often called 'work orders' and comprise all of the changes within a defined job, such as adding a new utility service or replacing part of a utility distribution network. The challenge of working with work orders is ensuring that the editor can seamlessly perform a complex set of edits to multiple datasets over an extended period of time without locking those datasets to other editors. ArcGIS addresses these workflows through a technology called versioning, which is available with ArcSDE geodatabases.

Versioning in the geodatabase

Versioning allows multiple users to edit the same data simultaneously by giving each editor their own 'version' of the original data to work with. Versioning works by recording any changes that are made to a dataset in an ArcSDE geodatabase. Only the changes made to the dataset are tracked and this avoids duplication of data.

Each editor works with their own version of the data until they decide to merge their changes back in with the original data. Once an individual editor's changes are incorporated into the original data, the edits are visible to other connected users.

The goal of versioning is to provide a large number of users with high-performance access to the geographic datasets in their geodatabase. Geographic information systems that employ versioning can efficiently support the use of datasets containing millions of rows accessed simultaneously by hundreds of users.

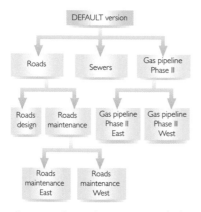

A conceptual view of versioning as applied to a public works department of a municipality

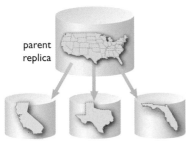

parent
replica

child replica   child replica   child replica

A conceptual view of replication showing the relationship between parent and child replicas.

Data distribution involves creating copies of data and dispensing it between two or more geodatabases. It allows two or more offices to be working on the same data in separate locations.

Data is distributed to improve data availability and performance by alleviating server contention and slow network access to a central server. This can help an organization balance the load on their geodatabases between users performing edits and those accessing it for reading operations.

Distributing data is often a requirement for mobile users or contractors who need to take part of their geodatabase into the field to edit, disconnecting from the network entirely for an indefinite amount of time. Replication makes this and many other workflows possible by extending the versioning framework of the geodatabase.

Replication with the geodatabase

Geodatabase replication is a data distribution method provided through ArcGIS. It is similar to other data distribution methods you are already familiar with, such as distribution of emails contained in an email repository or songs in a music library. Similar to adding and removing songs in your music library and then synchronizing your library with a portable music player, ArcGIS can add, modify, and delete features in a geodatabase and synchronize these changes with other geodatabases on mobile devices or other computers. Geodatabase replication offers a versatile solution to common workflows presented by GIS projects.

With geodatabase replication, data is distributed among different geodatabases by replicating all or part of the datasets within a central geodatabase. When a geodatabase is replicated, two replicas are created: one that resides in the original geodatabase and a related replica that is distributed to a different geodatabase. Any changes made to these replicas in their respective geodatabases can then be synchronized so that the data in one replica matches that in the related replica.

Geodatabase replication can be used in both connected and disconnected environments. In a connected environment, such as your office network, replicas are always accessible. In a disconnected environment, such as when you travel in the field away from your office, replicas are sporadically or possibly never connected on the same network.

Replication can work with local geodatabase connections to personal, file, and ArcSDE geodatabases. Replicas can also be connected over the internet using ArcGIS Server technology.

See the ArcGIS help system for more information on creating and synchronizing replicas.

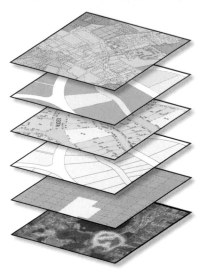

A GIS design is built around the set of thematic layers of information that will address a particular set of requirements. A thematic layer is a collection of common features, such as a road network, a collection of parcel boundaries, soil types, an elevation surface, satellite imagery for a certain date, and well locations.

Design starts with thematic layers

First, you define these thematic layers for your particular applications and information requirements. Then, you define each thematic layer in more detail. The characterization of each thematic layer will result in a specification of standard geodatabase data elements, such as feature classes, tables, relationships, raster datasets, subtypes, topologies, and domains.

When identifying thematic layers in your design, try to characterize each theme in terms of its visual representations, its expected usage in the GIS, its likely data sources, and its levels of resolution. These characteristics help describe the high-level contents expected from each theme.

Once you have identified the key thematic layers in a design, the next step is to develop specifications for representing the contents of each thematic layer in the physical database. These include how the geographic features are to be represented (for example, as points, lines, polygons, rasters, or tabular attributes); how the data is organized into feature classes, tables, and relationships; and how spatial and database integrity rules will be used to implement GIS behavior.

Layer	**Ownership parcels**
Map use	Parcels define landownership and are used for taxation.
Data source	Compiled from landownership transactions at local government .
Representation	Polygons in survey-aware feature classes and related annotation.
Spatial relationships	Parcel polygons cannot overlap and are covered by boundary lines.
Map scale and accuracy	Typical map scales are 1:1,200 and 1:2,400.
Symbology and annotation	Parcels often drawn using boundary features and related annotation.

Design steps

The eleven steps presented on the following page outline a general GIS database design process. The initial conceptual design steps help you identify and characterize each thematic layer. In the logical design phase, you begin to develop representation specifications, relationships, and, ultimately, geodatabase elements and their properties. In the physical design stage, you will test and refine your design through a series of initial implementations. You will also document your design.

Eleven steps to designing geodatabases

1 *Identify the information products that you will create and manage with your GIS*

Your GIS database design should reflect the work of your organization. Consider compiling and maintaining an inventory of map products, analytical models, Web mapping applications, data flows, database reports, key responsibilities, 3D views, and other mission-based requirements for your organization. List the data sources you currently use in this work. Use these to drive your data design needs.

2 *Identify the key data themes based on your information requirements*

Define more completely some of the key aspects of each data theme. Determine how each dataset will be used—for editing, for GIS modeling and analysis, representing your business workflows, and for mapping and 3D display. Specify the map use, the data sources, the spatial representations for each specified map scale; data accuracy and collection guidelines for each map view and 3D view; and how the theme is displayed, its symbology, text labels, and annotation.

3 *Specify the scale ranges and the spatial representations of each data theme at each scale*

Data is compiled for use at a specific range of map scales. Associate your geographic representation for each map scale. Geographic representation will often change between map scales (for example, from polygon to line or point). In many cases, you may need to generalize the feature representations for use at smaller scales. In other situations, you may need to collect alternative representations for different map scales.

4 *Decompose each representation into one or more geographic datasets*

Discrete features are modeled as feature classes of points, lines, and polygons. You can consider advanced data types such as topologies, networks, and terrains to model the relationships between elements in a layer as well as across datasets. For raster datasets, mosaics and catalog collections are options for managing very large collections. Surfaces can be modeled using features, such as contours, as well as using rasters and terrains.

5 *Define the tabular database structure and behavior for descriptive attributes*

Identify attribute fields and column types. Tables also might include attribute domains, relationships, and subtypes. Define any valid values, attribute ranges, and classifications (for use as domains). Use subtypes to control behaviors. Identify tabular relationships and associations for relationship classes.

6 *Define the spatial behavior, spatial relationships, and integrity rules for your datasets*

For features, you can add spatial behavior and capabilities and also characterize the spatial relationships inherent in your related features for a number of purposes using topologies, address locators, networks, terrains, and so on. For example, use topologies to model the spatial relationships of shared geometry and to enforce integrity rules. Use address locators to support geocoding. Use networks for tracing and pathfinding. For rasters, you can decide if you need a raster dataset or a raster catalog.

7 *Propose a geodatabase design*

Define the set of geodatabase elements you want in your design for each data theme. Study existing designs for ideas and approaches that work. Copy patterns and best practices from the ArcGIS Data Models.

8 *Design editing workflows and map display properties.*

Define the editing procedures and integrity rules (for example, all streets are split where they intersect other streets, and street segments connect at endpoints). Design editing workflows that help you to meet these integrity rules for your data. Define display properties for maps and 3D views. Determine the map display properties for each map scale. These will be used to define map layers.

9 *Assign responsibilities for building and maintaining each data layer.*

Determine who will be assigned the data maintenance work within your organization or assigned to other organizations. Understanding these roles is important. You will need to design how data conversion and transformation is used to import and export data across various partner organizations.

10 *Build a working prototype. Review and refine your design.*

Test your prototype design. Build a sample geodatabase copy of your proposed design using a file, personal, or ArcSDE Personal geodatabase. Build maps, run key applications, and perform editing operations to test the design's utility. Based on your prototype test results, revise and refine your design. Once you have a working schema, load a larger set of data (such as loading it into an ArcSDE geodatabase) to check out production, performance, scalability, and data management workflows.

11 *Document your geodatabase design.*

Various methods can be used to describe your database design and decisions. Use drawings, map layer examples, schema diagrams, simple reports, and metadata documents. Some users like using UML. However, UML cannot represent all the geographic properties and decisions to be made. Also, UML does not convey the key GIS design concepts such as thematic organization, topology rules, and network connectivity. Many users like using Visio® to create a graphic representation of their geodatabase schema such as those published with the ArcGIS data models. ESRI provides a tool that can help you capture these kinds of graphics of your data model elements using Visio.

Coordinate systems and map projections

2

ArcGIS software has the capability of spatially integrating map layers from datasets with many coordinate systems onto a map. This enables you to display, edit, and analyze map layers from many data sources, such as field survey data collected half a century ago, scanned mylar layers of national map series, imagery from aircraft and satellite sensors, and accurate field data collection with GPS. Spatial integration of data is done by defining a coordinate system for each geographic dataset, assigning a map projection to the map, and combining layers on the map. This chapter will help you select the right projection for your map's extent and use the correct coordinate system for your geographic datasets.

A geographic information system integrates map layers with different coordinate systems onto a common map frame. This is an important function for any GIS because geographers need to combine geographic data from many sources onto a map. All of the layers on a map need to be aligned, no matter what coordinate systems are used for the source geographic datasets.

This chapter explains these basic concepts to help you design maps and ensure that a map's layers are correctly aligned:

- Every map has a coordinate system which is selected for the geographic extent and purpose of the map. A map's coordinate system can be the same as the coordinate system used in your geographic datasets or it can be a specialized coordinate system for areas such as the world or continents.

- The process of projecting the earth's surface onto a flat map introduces distortions of various kinds. Through the selection of a map's coordinate system, you can control whether relative distances, areas, directions, or shapes are preserved.

- Every geographic dataset should have a well-defined coordinate system, which references the mathematical model of the earth's shape known as a datum. A geographic dataset stores positions using either spherical coordinates (latitude and longitude values on the earth) or projected coordinates (x and y distances from a point of origin on a projection). Each geographic dataset has a spatial reference, which defines the coordinate system along with other parameters for edit tolerances and resolution.

- ArcGIS will automatically align map layers from geographic datasets which have a well-defined coordinate system onto the coordinate system for the map. Aligning map layers lets you analyze the spatial relationships of entities from many different geographic datasets.

When you receive datasets from other sources, they will usually, but not always, have a well-defined coordinate system. You can diagnose and correct a missing coordinate system reference. The procedure to do so is discussed later in this chapter.

This exploded view of a map demonstrates how layers with distinct coordinate systems are spatially integrated in a GIS. Each of the four layers shown has three pieces of spatial information annotated beside it:

• The actual coordinate position of the identical place on each layer marked by the crosshair symbol.

• The type of coordinate system used by the geographic dataset—Geographic coordinates (latitude and longitude values), the State Plane Zone system of the United States, and the Universal Transverse Mercator system (UTM), used widely around the world.

• The type of datum, the mathematical model used for the earth's shape—the World Geodetic System of 1984, the North American Datum of 1983, and the North American Datum of 1927.

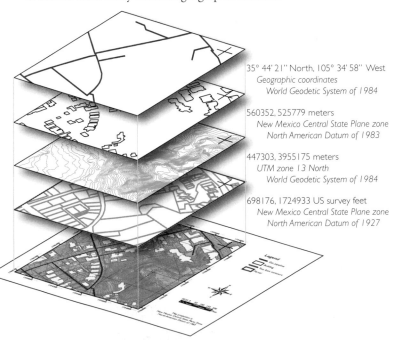

35° 44' 21'' North, 105° 34' 58'' West
Geographic coordinates
World Geodetic System of 1984

560352, 525779 meters
New Mexico Central State Plane zone
North American Datum of 1983

447303, 3955175 meters
UTM zone 13 North
World Geodetic System of 1984

698176, 1724933 US survey feet
New Mexico Central State Plane zone
North American Datum of 1927

This illustration shows a projection of India georeferenced from the earth's surface onto a map.

ArcGIS combines geographic data from many sources onto a common projected coordinate system. When each geographic dataset in a geodatabase has a well-defined coordinate system, map layers can be precisely aligned, spatial relationships can be uncovered, and geoprocessing operations such as overlay analysis are possible.

While geographic datasets commonly use geographic coordinates (latitude and longitude) for storing data, the only type of map that is truly based on geographic coordinates is a globe. All flat maps, whether printed on a sheet or displayed on a computer screen, involve a projected coordinate system.

The three map details below show examples of cylindrical, conic, and planar map projections.

Georeferencing on a map

All features on a map have a specific geographic location and extent. The ability to accurately describe geographic locations is critical in both mapping and GIS. This process is called georeferencing.

Describing the correct location and shape of features requires a framework for defining real-world locations. This is called a coordinate system and geographic datasets use one of two fundamental types of coordinate systems:

- Geographic coordinate systems (also called spherical coordinate systems) measure angles from the center of the earth relative to the equator and the prime meridian at Greenwich. Angles relative to the equator are latitude values and angles relative to the prime meridian are longitude values. These angles are measured in degrees, minutes, and seconds.

- Projected coordinate systems involve a projection of the earth's shape onto a flat surface. Some projections use a geometric shape which can be cut and unwrapped. The three main types of projected coordinate systems involve projections onto a cylinder, cone, and plane. Coordinates in projected coordinate systems are usually measured in meters or feet from a point of origin.

Map projections

Projected coordinate systems are based on map projections on a cylinder, cone, or plane. A map projection defines the mathematical transformation of portions of the earth's surface onto a map. A basic operation of a GIS is transforming coordinate systems from the source geodatabase dataset onto the projected coordinate system on a map using map projections.

This can be done either by the on-the-fly projection of layers to a map or by permanently transforming coordinates stored in a geographic dataset to another coordinate system. When you make maps, you can easily overlay geographic datasets from many different coordinate systems. When you perform some kinds of analysis or edit operations, then it may be necessary to ensure that related geographic datasets are transformed to a common coordinate system.

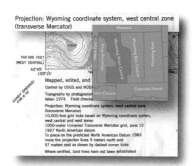

Transverse Mercator uses a cylindrical projection to map zones longer along the north–south axis.

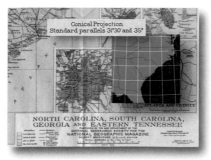

Conic projections are used for regional maps with regions longer in the east–west direction.

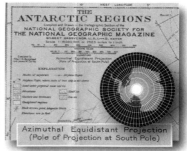

Planar projections are frequently used for mapping the polar regions of the world.

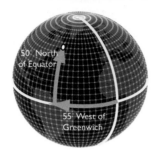

Spherical coordinates are angles measured from the center of the earth. Longitude angles are measured relative to the prime meridian that runs through the Greenwich Observatory in the United Kingdom. Latitude angles are measured relative to the equator.

In the geographic coordinate system, a point on the earth's surface is located by its longitude and latitude values, which are angular measurements from the earth's center to a point on the earth's surface relative to the prime meridian and the equator. Longitude and latitude values are represented with angular units of degrees, subdivided into 60 minutes and 60 seconds of arc.

Vertical lines in north–south directions are lines of equal longitude, or meridians. Horizontal lines in east–west directions are lines of equal latitude, or parallels. The meridians and parallels encompass the globe and form the graticular network.

The parallel midway between the poles is the equator and defines the line of zero latitude. The line of zero longitude is the prime meridian and by an international agreement in the late nineteenth century was defined to pass through the Greenwich Observatory in the United Kingdom.

Latitude and longitude values are traditionally represented either as decimal degrees or in degrees, minutes, and seconds. A degree has 60 minutes. A minute has 60 seconds. To convert latitude and longitude values formatted in degrees, minutes, and seconds to decimal degrees, apply this formula:

decimal degrees = degrees + (minutes / 60) + (seconds / 3600)

Latitude values are measured relative to the equator. Latitudes south of the equator have negative values and reach -90° at the South Pole. Latitudes north of the equator have positive values and reach 90° at the North Pole. Longitude values are measured relative to the prime meridian. Longitudes west of the prime meridian have negative values and reach -180° at the International Date Line in the Pacific Ocean. Longitudes east of the prime meridian have positive values and reach 180° at the International Date Line.

Although longitude and latitude values can locate precise positions on the surface of the globe, degree values do not represent uniform units of distance measure. Only along the equator does the distance represented by one degree of longitude approximate the distance represented by one degree of latitude.

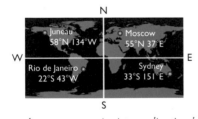

A common convention is to use directional suffixes for latitude and longitude values instead of positive and negative values. 'N' and 'S' mark north and south latitudes, 'E' and 'W' mark longitudes east and west of the prime meridian. This system divides the world into four quadrants. The longitude and latitude values for four cities in each quadrant—Juneau, Moscow, Rio de Janeiro, and Sydney—are shown in this form.

Above and below the equator, parallels of latitude become smaller until they become a single point at the North and South Poles where the meridians converge. As the meridians converge toward the poles, the distance represented by one degree of longitude decreases to zero. At the equator, one degree of longitude at the equator is approximately 111.3 kilometers, while at 60° latitude it is only 55.8 kilometers.

An error sometimes made by people new to GIS is to display geographic datasets that have latitude and longitude coordinates without a map projection. While it's easy to make a map in this way, this is a poor cartographic practice. When you display longitude and latitude values as unprojected x- and y-coordinates on a map, shapes of geographic features become progressively distorted at high latitudes and it's impossible to measure distances or areas accurately. A map scale is not meaningful for this kind of map because distances represented by degrees vary considerably with latitude.

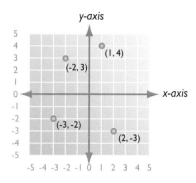

In a projected coordinate system, locations are identified by x, y coordinates on a grid, with the origin at the center of the grid. Each position has two values that reference it to that central location. One specifies its horizontal position and the other its vertical position. The two values are called the x-coordinate and y-coordinate. Using this notation, the coordinates at the origin are x = 0 and y = 0.

Cylindrical or conic projections are logically cut along a meridian or line of latitude such as the equator. Positions in the coordinate system are specified with x and y values relative to a selected point of origin.

A projected coordinate system is defined on a flat, two-dimensional surface and can display geographic features with constant lengths, directions, or areas across two dimensions. Like geographic coordinate systems, a projected coordinate system is always based on a sphere or spheroid that is a model of the earth's shape.

Properties of projections

A map is the result of a transformation from the earth's three-dimensional shape onto a flat sheet. This transformation is called a map projection. Most map projections are based on projecting the earth's surface onto either a plane, cone, or cylinder, or some combination of the three. Cones and cylinders are used because once the earth's surface is projected, these shapes can be cut and flattened to a flat sheet without further distortion.

Any mathematical projection from a sphere (or spheroid) onto a flat map involves some type of distortion. Different projections cause different types of distortions. Some projections are designed to minimize the distortion of one or two of the data's characteristics. A projection could maintain the area of a feature but alter its shape. Many projections involve a compromise among these types of distortion.

In summary, these are the classes of projections and the properties they are designed to maintain.

- Conformal projections preserve local shape. To preserve individual angles describing the spatial relationships, a conformal projection must show the perpendicular graticule lines intersecting at 90-degree angles on the map. A map projection accomplishes this by maintaining all angles. The drawback is that the area enclosed by a series of arcs may be greatly distorted in the process.

- Equal area projections preserve the area of displayed features. To do this, other properties such as shape, angle, and scale are distorted. In equal area projections, the meridians and parallels may not intersect at right angles. In some instances, especially maps of smaller regions, shapes are not obviously distorted, and distinguishing an equal area projection from a conformal projection is difficult unless documented or measured.

- Equidistant projections preserve the distances between certain points. Scale is not maintained correctly by any projection throughout an entire map. Most equidistant projections have one or more lines in which the length of the line on a map is the same length (at map scale) as the same line on the globe, regardless of whether it is a great or small circle, or straight or curved. Such distances are said to be true. Keep in mind that no projection is equidistant to and from all points on a map.

- True-direction projections maintain the shortest route between two points on a curved surface such as a sphere. This is the great circle on which the two points lie. True-direction, or Azimuthal projections, maintain some of the great circle arcs, giving the directions or azimuths of all points on the map correctly with respect to the center. Some true-direction projections are also conformal, equal area, or equidistant.

Chapter 2 — Coordinate systems and map projections   25

The fundamental geographic points and lines on the earth are the North and South Poles, the Equator, the tropics of Cancer and Capricorn, and the Arctic and Antarctic Circles. While the prime meridian could have been defined at any longitude, zero degrees longitude passes through the Greenwich Observatory in the United Kingdom by international agreement.

The Arctic and Antarctic Circles mark the most northerly and southerly latitudes for which the sun never sets on the summer solstice and never rises on the winter solstice.

The Tropic of Cancer and Capricorn mark the extent of the tropical range of latitude within which the sun can occupy the zenith position in the sky.

North pole *latitude is 90° North*
Arctic Circle
latitude is 66° 33' 39" North
Tropic of Cancer
latitude is 23° 26' 21" North
Prime Meridian
longitude is 0°
Equator
latitude is 0°
Tropic of Capricorn
latitude is 23° 26' 21" South
latitude is 66° 33' 39" South
Antarctic Circle
South pole *latitude is 90° South*

All coordinate systems used for geography are based on a mathematical model of the earth's surface.

We tend to think of the earth's surface as a sphere and it is nearly so. A sphere is the shape that a large body in space will assume due to the influence of gravity. But because of the earth's rotation, the earth is slightly wider along the equator. The radius of the earth from the center of the earth to any point on the equator is slightly more than the radius from the earth's center to either pole.

This difference in radius is about 1/3 of a percent. That is, the radius from earth's center to either pole is about 99 2/3 percent of the distance from the earth's center to any point on the equator.

The shape and size model of the earth is the basis for all coordinate systems used for geography and is called the spheroid.

Definition of a spheroid

A sphere is based on rotating a circle and a spheroid (or ellipsoid) is based on rotating an ellipse. The shape of an ellipse can be defined by two radii. The longer radius is called the semimajor axis, and the shorter radius is called the semiminor axis.

Over time, the earth has been surveyed many times. These surveys have resulted in many spheroids that represent the earth.

In the past, a spheroid was chosen to fit one country or a particular area. A spheroid that best fits one region is not necessarily the same one that fits another region. Until several decades ago, North American geographic data used a spheroid determined by Clarke in 1866.

Today, spheroids are computed using satellites and are considered to be very accurate. The two spheroids used today are the Geodetic Reference System of 1980 (GRS80) and the World Geodetic System of 1984 (WGS84).

For comparison, these are the radius distances and inverse of flattening values of three commonly used spheroids. For practical purposes, GRS80 and WGS84 are considered identical.

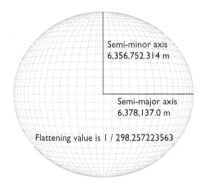

Semi-minor axis
6,356,752.314 m

Semi-major axis
6,378,137.0 m

Flattening value is 1 / 298.257223563

The current standard for spheroids is WGS84 which has these values for the earth's spheroid.

A spheroid has a flattening value that is the ratio of the difference in length between the two axes versus the equatorial axis length. The flattening value is about .00335281, close to one-third of a percent.

Spheroids	Semi-major axis	Semi-minor axis	Inverse of flattening
Clarke 1866	6378206.4 m	6356583.8 m	294.9786982
GRS80	6378137	6356752.31414	298.257222101
WGS84	6378137	6356752.31424518	298.257223563

DATUMS

Whenever you change the datum, or more correctly, the geographic coordinate system, the coordinate values of your data will change. Here are the latitude and longitude coordinates in degrees, minutes, and seconds of a control point in Redlands, California, on the North American Datum of 1983 (NAD83).

34° 01' 43.77884", -117° 12' 57.75961"

Here's the same point on the North American Datum of 1927 (NAD27).

34° 01' 43.72995", -117° 12' 54.61539"

The longitude value differs by approximately three seconds, while the latitude value differs by about 0.05 seconds.

Here are the coordinates for the same control point based upon WGS 1984.

34° 01' 43.79664", -117° 12' 57.80264"

All coordinate systems used for map making, whether geographic coordinate systems (latitude and longitude) or projected coordinate systems (such as UTM or State Plane Coordinate System) are based on a datum.

A datum provides a frame of reference for measuring locations on the surface of the earth. It defines the origin and orientation of latitude and longitude lines. A way to explain it is that a datum defines the position of the spheroid relative to the center of the earth while a spheroid approximates the shape of the earth.

As a GIS professional, you should be aware of the major datums in use such as NAD27, ED50, and WGS84. Understanding datums is important for precisely aligning geographic data. When you add geographic data to a map and discover subtle offsets in the alignment of your map layers, a possible cause is that one or more of your geographic data sources might have an incorrectly defined datum.

Geocentric datums

In the last several decades, satellite data has provided geodesists with new measurements to define the best earth-fitting spheroid, which relates coordinates to the earth's center of mass. An earth-centered, or geocentric, datum uses the earth's center of mass as the origin. One of the first geocentric datums was WGS 1984. Many countries now use geocentric datums.

Local datums

A local datum aligns its spheroid to closely fit the earth's surface in a particular area. A point on the surface of the spheroid is matched to a particular position on the surface of the earth. This point is known as the origin point of the datum. The coordinates of the origin point are fixed, and all other points are calculated from it.

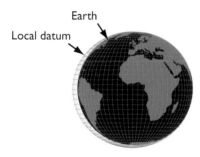

A conceptual (and exaggerated) view of the offset between a local datum and the earth.

The coordinate system origin of a local datum is not at the center of the earth, but offset from the earth's center. NAD 1927 and the European Datum of 1950 (ED 1950) are local datums. NAD 1927 is designed to fit North America reasonably well, while ED 1950 was created for use in Europe. Because a local datum aligns its spheroid so closely to a particular area on the earth's surface, it's not suitable for use outside the area for which it was designed.

North American datums

There was an ongoing effort at the state level in the U.S. to readjust the NAD 1983 datum to a higher level of accuracy using state-of-the-art surveying techniques that were not widely available when the NAD 1983 datum was being developed. This effort, known as the High Accuracy Reference Network (HARN), or High Precision Geodetic Network (HPGN), was a cooperative project between the National Geodetic Survey and the individual states.

A nationwide readjustment, called NAD83 (NSRS2007) was conducted recently.

The two horizontal datums used almost exclusively in North America are NAD 1927 and NAD 1983. NAD 1927 uses the Clarke 1866 spheroid to represent the shape of the earth. The origin of this datum is a point on the earth referred to as Meades Ranch in Kansas. Many NAD 1927 control points were calculated from observations starting in the 1800s.

The North American Datum of 1983 was based on both earth and satellite observations. The origin for this datum was the earth's center of mass. This affected the surface location of all longitude–latitude values enough to cause locations of previous control points in North America to shift, sometimes as much as 500 feet.

Maps are drawn with a map projection. Positions for geographic data is stored using a coordinate system based on a map projection. It's important to understand that there is no requirement that the map projection of the coordinates in a geographic dataset match the map projection of the map that you make. As shown at the beginning of this chapter, a map can be drawn with any map projection suitable for a map's purpose and extent from geographic data.

The geodatabase stores information about coordinate systems of each geographic dataset with spatial references, a collection of properties that describe the coordinate system as well as other properties such as the coordinate resolution for x,y coordinates and optional z- and m- (measure) coordinates. Spatial references allow data from many sources and coordinate systems to be displayed together and to allow either coordinate transformation when necessary for editing and analysis or map layer projections for publication.

In a geodatabase, a spatial reference has two main components: a coordinate system and a spatial reference domain. The coordinate system defines where feature coordinates are located on the earth, while the spatial reference domain controls how those coordinates are stored in the geodatabase and processed in ArcGIS.

A spatial reference describes where features are located in the real world. You define a spatial reference when creating a geodatabase feature dataset or stand-alone feature class. The spatial reference includes a coordinate system for x-, y-, and z-values as well as tolerance and resolution values for x-, y-, z-, and m-values.

x, y coordinates are georeferenced with either a geographic or projected coordinate system. A geographic coordinate system (GCS) is defined by a datum, an angular unit of measure (usually degrees), and a prime meridian. A projected coordinate system (PCS) consists of a linear unit of measure (usually meters or feet), a map projection, the specific parameters used by the map projection, and a geographic coordinate system.

A projected or geographic coordinate system can have a vertical coordinate system (VCS) as an optional property. A VCS georeferences z-values, most commonly used to denote elevation. A vertical coordinate system includes a geodetic or vertical datum, a linear unit of measure, an axis direction, and a vertical shift.

All the feature classes in a feature dataset share the same spatial reference. This enables topological editing and the building of networks and fabrics. Feature classes not in a feature dataset have their own spatial reference. For raster data, the spatial reference describes the coordinate system associated with the data after an image-to-map transformation has been applied.

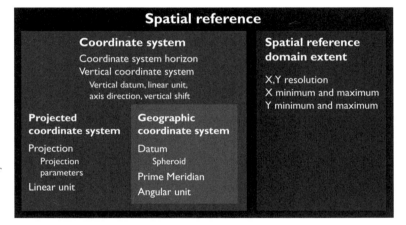

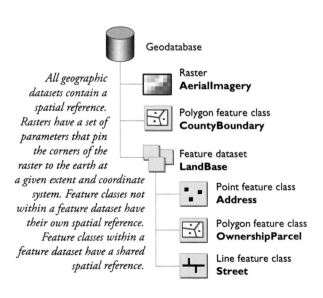

All geographic datasets contain a spatial reference. Rasters have a set of parameters that pin the corners of the raster to the earth at a given extent and coordinate system. Feature classes not within a feature dataset have their own spatial reference. Feature classes within a feature dataset have a shared spatial reference.

An organization doing GIS is combining many datasets from different sources. For datasets from outside your organization, unless you need to edit them together or add structure such as a network, leave their spatial reference as is.

Datasets created and managed within your organization should share a common spatial reference, be compatible with other local organizations, and use WGS84 or other recent spheroid.

The way that ArcGIS spatially integrates layers on a map is by spatial references, a series of parameters that define the coordinate system and other properties for each geographic dataset in the geodatabase.

An ArcGIS spatial reference includes settings for:

- The coordinate system and parameters.
- The coordinate precision with which coordinates are stored.
- Processing tolerances (the XY, Z, and M tolerances) which determines whether two close points should be considered coincident.
- The spatial domain (or map extent) covered by the dataset.

Each coordinate system of a spatial reference is defined by:

- Its measurement framework which is either geographic (spherical coordinates measured from the earth's center) or planimetric (coordinates projected onto a two-dimensional plane).
- Unit of measurement, typically feet or meters for projected coordinate systems or decimal degrees for latitude–longitude.
- Other measurement system properties such as spheroid of reference, datum, and projection parameters like standard parallels, a central meridian, and possible x- and y-direction shifts.

properties of a spatial reference

a spatial reference has a spherical or projected **coordinate system.** Latitude and longitude values are spherical coordinates and not a map projection. A map projection is usually on a cone, cylinder, or plane, sometimes with modifications

the **coordinate resolution** defines a fine map grid on which the x- and y-coordinates of every point and vertex will snap to

coordinate precision

the **XY tolerance** determines whether two points are joined during a geoprocessing operation

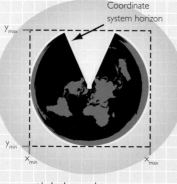

Coordinate system horizon

y_{max}

y_{min}

x_{min} x_{max}

the **spatial domain** consists of the range of x and y values, further restricted by the coordinate system horizon

Lines also have an optional m-value, which stores linear measures along a route.

m_0 m_1 m_2 m_3 m_4 m_5

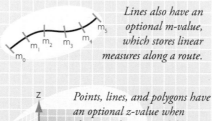

z

Points, lines, and polygons have an optional z-value when elevation data is available.

x

y

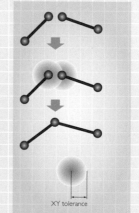

XY tolerance

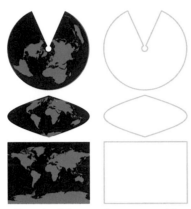

Examples of three coordinate systems, Albers Equal Area Conic, Sinusoidal, and Miller Cylindrical, and their respective coordinate system horizons.

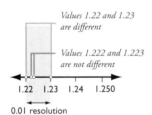

To understand coordinate resolution, consider this example: If a spatial reference has an XY resolution of 0.01, then x-coordinates 1.22 and 1.23 are different, but 1.222 and 1.223 are not.

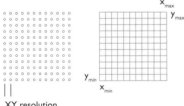

XY resolution

The spatial reference domain is influenced by XY resolution. As XY resolution gets larger (coarser), the maximum x, y values become larger. Therefore, a spatial reference domain extent with a large (coarse) XY resolution contains more x,y coordinate area than one with a smaller (finer) XY resolution.

With double-precision coordinates in the geodatabase, the XY resolution of the spatial reference domain can be very fine. With 15 places of significant digits, the entire world can be mapped at an XY resolution of a millimeter.

Coordinate system horizon

Every geographic coordinate system and projected coordinate system contains a valid area known as the coordinate system horizon, which contains the set of legal x and y values in a coordinate system. While editing, you will not be able to add any features outside the coordinate system horizon. Some coordinate system horizons cover the earth while some do not.

Spatial reference domain

The spatial reference domain is the allowable set of values for x, y, z, and m in a feature class. X, y values define a geographic position in two-dimensional space; z- and m-values are optional and typically used for elevation and route measurements.

The spatial reference domain also defines how coordinates in a feature class are managed in a geodatabase. First, it establishes the number of significant decimal places for the coordinates that comprise the features in a feature class. Second, it provides an areal extent of x, y coordinate space and a set of possible x, y coordinate locations inside that space. Third, it determines the behavior of coordinates that are very close together during relational and topological processing operations. These three aspects of the spatial reference domain are closely related—a change in one will affect the others.

Coordinate resolution

Coordinate resolution is the numeric precision that defines the number of decimal places or significant digits used to store feature coordinates. It is the minimum distance that separates unique x, y, z-, or m-values. You can set three types of coordinate resolution for a spatial reference:

1. XY resolution for x, y, coordinates.

2. Z resolution for z-values

3. M resolution for m-values

XY and Z resolution values are in the same units as the associated coordinate system. For example, if a spatial reference uses a projected coordinate system with units of meters, the XY and Z resolution values are defined in meters. M resolution values can have different units than the associated coordinate system.

Coordinate grid

Every feature class in a geodatabase has a spatial reference, which in turn has a coordinate grid. The structure of the grid is a mesh of points that forms a square in two-dimensional space with an origin point and an XY resolution. The coordinate grid represents the x, y coordinate space that can be stored within a feature class.

Associated with the grid is its XY resolution, the smallest distance between different x-coordinates or different y-coordinates. Adjacent horizontal and vertical mesh points in the coordinate grid are separated by a distance equal to the XY resolution.

Spatial reference domain extent
Xmin = -5,136,135 m
Xmax = 14,890, 535 m
Ymin = -9, 950,490 m
Ymax = 9,992,387 m
XY resolution = 2.2 x 10-9 m

Spatial reference domain extent
Xmin = -21,059,800 m
Xmax = 21,059,800 m
Ymin = -10,226,500 m
Ymax = 31,893,100 m
XY resolution = 4.7 x 10-9 m

Feature locations at A, B, and C are snapped to the closest coordinate grid location before any geographic processing.

Spatial reference domain extent

The spatial reference domain extent is the maximum allowable area (in terms of x,y coordinate values) that is covered by the coordinate grid for a given spatial reference.

Given a user-specified coordinate system and XY resolution, ArcGIS will automatically determine a spatial reference domain extent that includes the coordinate system horizon.

At left are two coordinate systems with their coordinate system horizons and spatial reference domain extent. The XY domain extent must be square.

When a feature class is processed in ArcGIS with geometric, relational, or topological operations, its feature coordinate values must be referenced with respect to the coordinate grid. X, y coordinates must lie exactly on the intersections of the vertical and horizontal lines (respectively) of the coordinate grid.

Feature coordinates that are not situated at the mesh points of the grid will be moved ("snapped") to the nearest mesh points before they can be used. This movement causes values to be rounded to intervals of the XY resolution. Feature coordinates that are aligned to the coordinate grid are called snapped coordinates.

Coordinate tolerances

The coordinate tolerances defines the minimal distance that a feature coordinate is allowed to move during relational and topological operations. Recall that three types of tolerances are set in a geodatabase: XY tolerance, Z tolerance, and M tolerance.

XY tolerance is the most commonly used. It is an extremely small distance that determines whether two (or more vertices) are close enough to be given the same x, y coordinate value or if they are far enough apart to each have their own x, y coordinate values. It also defines the distance that coordinates can move in x, y coordinate space during clustering operations (topology validation).

Different tolerance values will produce different results for relational and topological operations. For example, two geometries might be classified as disjoint with a small XY tolerance, but a larger XY tolerance may classify them as touching.

Generally, XY tolerance should be an order of magnitude larger than the XY resolution. This will help ensure data is not accidentally corrupted during processing. XY tolerance is not intended to be used for generalizing geometry shapes but instead to be used for integrating points, lines, and boundaries during relational and topological operations.

These same concepts for XY tolerance apply to Z and M tolerances. If z-values are applied, then two points in space can have the same x and y coordinate values, but different z-values (relative to the Z tolerance), and be different points.

All positions in geographic datasets, whether vector data in a feature class or cell data in a raster, are stored using either a geographic coordinate system (latitude and longitude values) or a projected coordinate system (x- and y-coordinates usually using length units of meters or feet).

It's common and acceptable to store geographic data using latitude and longitude values, but it's rarely acceptable to display this data on a map without applying a map projection.

The Universal Transverse Mercator system is commonly used for geographic datasets covering an area larger than the zones in a national mapping system, such as the U.S. State Plane Coordinate System. UTM is popular because it spans the globe and can be used worldwide. Some regions, particularly Europe, prefer to use a similar system called Gauss-Krüger.

If your geographic data is stored in a projected coordinate system, it should usually be in a coordinate system compatible with local mapping standards. In the United States, the two coordinate systems most commonly used are the State Plane Coordinate System (SPCS) and the Universal Transverse Mercator system (UTM). The choice will be guided by whether the area you are mapping fits within one State Plane zone or if a larger UTM zone is required. Map agencies for other countries around the world have standard coordinate systems that you should most likely use.

Remember that with ArcGIS, you can apply a map projection suited for the map use that is different from the map projection used for the coordinate system of the geographic datasets.

Troubleshooting unknown coordinate systems

When you import geographic datasets from other sources they usually have well-defined coordinate systems, but not always. Below are a few tips to identify the correct coordinate system. Once the correct coordinate system is identified, you can assign this to the spatial reference of the dataset.

In ArcCatalog, select a geographic dataset and with the Properties dialog box examine the coordinate system and extent. If the coordinates have values between -180 and +180 in the x-direction and -90 and +90 in the y-direction, then you likely have geographic coordinates. The next step is to identify the datum used for the data. Globally, WGS84 is the current standard datum. In the U.S., NAD 1927 is common on old datasets and NAD 1983 is common on new U.S. datasets.

If the data shows an extent in which the coordinates to the left of the decimal are 6, 7, or 8 digits, the data is probably in a projected zone. Try UTM first and identify the zone with the chart at the end of this chapter.

If UTM doesn't work, then try whatever coordinate systems are in use locally, such as a national projection. If you are in the US, try the local State Plane Coordinate System zone. See the chart later in this chapter. As with UTM, you need to determine whether the WGS84, NAD27, or NAD83 datums were applied.

If you observe offsets from a few meters to several hundred meters, then you are either seeing inaccuracies inherent in the data or possibly a datum shift. In the U.S., data in the UTM coordinate system referenced to the NAD83 datum is approximately 200 meters north of the same data referenced instead to the NAD27 datum. If you see an approximately 200-meter difference along the north-south axis, that is diagnostic of an incorrect datum assignment.

Guidelines for selecting a coordinate system for your spatial reference

	Datasets spanning the world or large regions	Datasets spanning countries and regions	Datasets spanning local areas

Geographic coordinates

Geographic coordinates are the best choice of a coordinate system for data covering the world or large regions. Although maps of continents commonly use conic or other projections, geographic coordinates are preferred because of their widespread usage.

Use geographic coordinates if your datasets are built from GPS data or when the area covered by a dataset is larger than a UTM zone. UTM zones are 6 degrees wide and can be extended up to 12 degrees.

Use geographic coordinates if your data is built from data collected by GPS receivers. Despite the fact that geographic coordinates span the world, your dataset can maintain very high precision, as explained below.

ArcMap performs on-the-fly projection of all map layers from any geographic dataset onto the projected coordinate system of your map. Any coordinate system that is reasonable for a dataset's extent will work. One factor that will influence your choice of coordinate system is whether you edit datasets together. If you do, then they will need to share the same spatial reference.

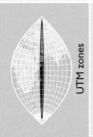

UTM zones

Use a UTM zone if the area covered by your dataset is too large for your national map coordinate system and is within a span of 12 degrees of longitude. The Universal Transverse Mercator coordinate system preserves local angles and shapes, although there is some distortion of distance and direction.

Use a UTM zone if the area covered by your dataset is larger than the zone specified by your national mapping system, if you exchange data with organizations using this UTM zone, or if no national map coordinate system exists in the area. UTM is widely used throughout the world.

Coordinate precision in a spatial reference

as small as 10^{-9} meters

GIS professionals want to know that coordinates in any coordinate system are stored with sufficient precision. The geodatabase uses high-precision storage of coordinates and any spatial reference can map the entire world with a precision of up to 10 nanometers. This extremely high resolution is seldom applied in practice; usually you can use the default setting which also has very high resolution.

National map system

Many countries in the world have a national map coordinate system, usually with multiple zones. Use your national map coordinate zones for compliance with government agencies and data exchange with local organizations.

When a spatial reference is created with a geographic coordinate system using degrees as units, the default resolution is 10^{-9} degrees, or approximately 0.11 millimeters. When a spatial reference is created with any projected coordinate system using meters as units, the default resolution is 0.0001 meters, or 0.1 millimeters.

All coordinate systems, whether projected or geographic, are based on a datum. A datum is defined on an ellipsoid, a shape model of the earth with semi-minor and semi-major radius lengths of the earth.

Choosing a datum for new geographic datasets

For all new geographic datasets, you should use the current datum that is in use in your country or organization, especially if you are performing any coordinate processing with geodetic control points. In the U.S., two datums in current use are NAD83 (CORS96) and NAD83 (NSRS2007).

If you are collecting coordinates with GPS equipment and not checking positions against geodetic control points, then use WGS84. Unlike early datums which are based on a spheroid fitted to a continent such as the North American Datum of 1927 (NAD27), WGS84 is a geocentric datum with its center precisely located at the center of the earth.

Handling datums from older geographic datasets

If you are working with or importing geographic datasets built some time ago, you will encounter earlier datums such as NAD27 or the European Datum of 1950 (ED50).

ArcGIS will correctly project coordinate values from these datums, provided they are properly defined in the spatial reference and an appropriate geographic (datum) transformation is set.

If these older geographic datasets are for static display and are not updated, then you can keep these datasets defined with the older datums. If you are adding new data to these datasets, such as GPS points, or editing this data together with other datasets, then you will want to transform the coordinate system to another coordinate system based on a modern datum such as WGS84.

The Universal Transverse Mercator system is primarily based on a set of 60 zones of a tranverse cylindrical projection with secant lines to minimize distortion within each zone.

The Universal Transverse Mercator (UTM) system is widely used throughout the world. The UTM system was established soon after World War II because of the U.S. military need for a standardized spatial coordinate system for the world. Since then, UTM has become popular because of ease of use and world-wide coverage. Global Positioning System (GPS) receivers often display UTM coordinates as well as latitude and longitude values.

Projection method

The UTM system uses the Transverse Mercator projection with two secant lines parallel to and approximately 180 kilometers to each side of the central meridian of the UTM zone. The original UTM system contained a specification for a grid-based location system, but contemporary use of the UTM system emphasizes the use of coordinate locations within each of the 60 zones instead of the grid-based location system.

The origin for each UTM zone is a central meridian and the equator. The central meridian for each zone is 6 degrees of longitude offset from the adjoining zone.

To eliminate negative coordinates, the UTM coordinate system alters the coordinate values at the origin. The value given to the central meridian is the false easting, and the value assigned to the equator is the false northing. A false easting of 500,000 meters is applied. A north zone has a false northing of zero, while a south zone has a false northing of 10,000,000 meters.

Properties

The UTM system uses the transverse cylindrical projection which is conformal with accurate representation of small shapes. There is minimal distortion of larger shapes within each zone. Areas of shapes have minimal distortion within each UTM zone and local angles are true.

The map scale within a UTM zone is constant along the central meridian but a scale factor of 0.9996 is applied to reduce lateral distortion within each zone. With this scale factor, lines lying 180 kilometers east and west of and parallel to the central meridian have a scale factor of one.

Using the UTM system

The Universal Transverse Mercator system is designed for a scale error not exceeding 1 part in 2500 within each zone. For some applications, a map area might extend a UTM zone, and when this is done, error and distortion increase. It is not appropriate to use UTM for areas that span more than a few zones and the extent should be limited to 9° on both sides of the central meridian. Beyond that range, data projected to the Transverse Mercator projection may not project back to the same position.

UTM is used for United States Geologic Survey topographic quadrangles at 1:100,000 scale. Many counties in the U.S. use local UTM zones based on the official geographic coordinate systems in use.

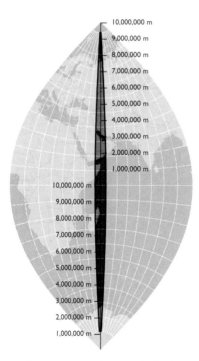

Each UTM zone is 6 degrees wide, divided into north and south projections. This is zone 39. Recall that the original definition of a meter is 1/10,000,000 of the distance along the spheroid from the equator to the pole. Each UTM zone has a north—south range of roughly 10,000 kilometers, minus the polar regions, and about 668 kilometers wide at the equator.

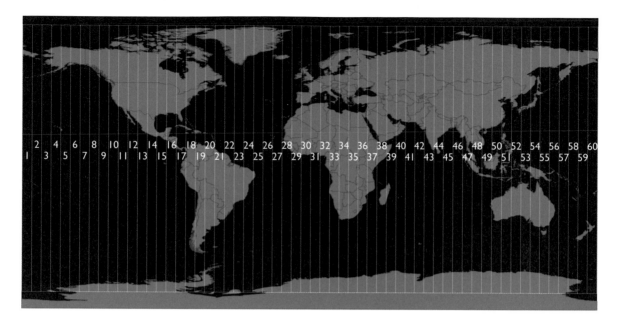

This graphic shows the generalized UTM zones as defined in ArcGIS. In UTM, there are several wider UTM zones that apply to the country of Norway but do not appear in this map or table.

UTM is a specialized application of the Transverse Mercator projection. The globe is divided into 60 north and south zones, each spanning 6° of longitude. Each zone has its own central meridian. Zones 1N and 1S start at 180° W. The limits of each zone are 84° N and 80° S, with the division between north and south zones occurring at the equator. Additional zones exist for the polar regions and use the Universal Polar Stereographic coordinate system.

Zone	Longitude range	Meridian
1	180°W—174°W	177°W
2	174°W—168°W	171°W
3	168°W—162°W	165°W
4	162°W—156°W	159°W
5	156°W—150°W	153°W
6	150°W—144°W	147°W
7	144°W—138°W	141°W
8	138°W—132°W	135°W
9	132°W—126°W	129°W
10	126°W—120°W	123°W
11	120°W—114°W	117°W
12	114°W—108°W	111°W
13	108°W—102°W	105°W
14	102°W—96°W	99°W
15	96°W—90°W	93°W
16	90°W—84°W	87°W
17	84°W—78°W	81°W
18	78°W—72°W	75°W
19	72°W—66°W	69°W
20	66°W—60°W	63°W

Zone	Longitude range	Meridian
21	60°W—54°W	57°W
22	54°W—48°W	51°W
23	48°W—42°W	45°W
24	42°W—36°W	39°W
25	36°W—30°W	33°W
26	30°W—24°W	27°W
27	24°W—18°W	21°W
28	18°W—12°W	15°W
29	12°W—6°W	9°W
30	6°W—0°	3°W
31	0°—6°E	3°E
32	6°E—12°E	9°E
33	12°E—18°E	15°E
34	18°E—24°E	21°E
35	24°E—30°E	27°E
36	30°E—36°E	33°E
37	36°E—42°E	39°E
38	42°E—48°E	45°E
39	48°E—54°E	51°E
40	54°E—60°E	57°E

Zone	Longitude range	Meridian
41	60°E—66°E	63°E
42	66°E—72°E	69°E
43	72°E—78°E	75°E
44	78°E—84°E	81°E
45	84°E—90°E	87°E
46	90°E—96°E	93°E
47	96°E—102°E	99°E
48	102°E—108°E	105°E
49	108°E—114°E	111°E
50	114°E—120°E	117°E
51	120°E—126°E	123°E
52	126°E—132°E	129°E
53	132°E—138°E	135°E
54	138°E—144°E	141°E
55	144°E—150°E	147°E
56	150°E—156°E	153°E
57	156°E—162°E	159°E
58	162°E—168°E	165°E
59	168°E—174°E	171°E
60	174°E—180°E	177°E

The State Plane Coordinate System (SPCS) divides the 50 United States along with Puerto Rico and the U.S. Virgin Islands into more than 120 sections, referred to as state plane zones. Each zone has a name such as "Oregon North" with a projection type and parameters optimized for mapping that region.

Many government agencies and organizations under contract at the state, county, and municipal levels use the State Plane Coordinate System. These organizations use SPCS because it is a long established standard and suitable for mapping at large scale mapping.

The State Plane Coordinate System was designed for large-scale mapping in the United States in the 1930s by the U.S. Coast and Geodetic Survey to provide a common reference system for surveyors and mappers. The goal was to design a conformal mapping system spanning the United States with a maximum scale distortion of one part in 10,000, which at the time was considered the limit of surveying accuracy.

State Plane Coordinate System projections

Three conformal projections were chosen: the Lambert Conformal Conic for states that are longer east–west, such as Tennessee and Kentucky; the Transverse Mercator projection for states that are longer north–south, such as Illinois and Vermont; and the Oblique Mercator projection for the panhandle of Alaska, because it lays at an angle.

To maintain an accuracy of one part in 10,000, it was necessary to divide many states into zones. Each zone has its own central meridian or standard parallels to maintain the desired level of accuracy. The boundaries of these zones always follow county boundaries. Smaller states, such as Connecticut, require only one zone, while the largest state, Alaska, is composed of 10 zones and uses all three projections.

State Plane Coordinate Systems and the North American Datum

There are two versions of the State Plane Coordinate System based on the NAD27 and NAD83 datums called SPCS 27 and SPCS 83, respectively. Technological advancements of the last 50 years have led to improvements in the measurement of distances, angles, and the earth's size and shape. NAD27 has its datum originate at Meades Ranch in Kansas and NAD83 shifted its datum origin to the earth's center of mass for compatibility with satellite systems. This shift made it necessary to redefine SPCS 27 and the redefined and updated system is called the State Plane Coordinate System of 1983 (SPCS 83).

The coordinates for points are different for SPCS 27 and SPCS 83. There are several reasons for this. For SPCS 83, all State Plane coordinates are in units of meters (instead of feet as with SPCS 27), the shape of the spheroid of the earth is slightly different, and some states have changed the definition and shapes of their zones.

Hawaii Zone 5

Hawaii Zone 4

Hawaii Zone 3

Hawaii Zone 2

Hawaii Zone 1

Alaska Zone 9 Alaska Zone 8
Alaska Zone 7
Alaska Zone 6
Alaska Zone 5
Alaska Zone 4
Alaska Zone 3
Alaska Zone 2

Alaska Zone 1

10

Alaska Zone 1 is unique among all state plane zones. The panhandle of Alaska lays at an angle, so conventional cylindrical, conic, or planar projections are not sufficient to control projection distortions. The solution is a tilted cylindrical projection, the Oblique Mercator.

All but one of the state plane zones has either a Transverse Mercator projection or a Lambert Conformal Conic projection. Each projection is optimized for wide or narrow zones.

States or zones that are wider in the east-wide direction use the Lambert Conformal Conic projection. A conic projection has no distortion along it's line of latitude.

Tennessee has a single zone with the Lambert Conformal Conic projection.

States or zones that are longer in the north—south axis use the Transverse Mercator projection. A cylindrical projection is distortion-free along it's contact meridian lines.

Vermont has a single zone with the Transverse Mercator projection.

Choosing coordinate systems

In the United States, the two most common projected coordinate systems are the State Plane Coordinate System (SPCS) and the Universal Transverse Mercator (UTM) system. Which is best depends on several factors.

Use SPCS if you frequently exchange geographic data with state and local agencies that use SPCS.

Another advantage of SPCS is that scale distortions at the edge of the zone will not exceed 1:10,000, in contrast to UTM's allowable distortion, 1:2,500. This means that a mile mapped at the edge of a state plane zone will have maximum distortion of 5,280 feet / 10,000, or half a foot.

Use UTM if your map area is larger than a state plane zone. UTM zones are 6 degrees of longitude wide, about 667 km (414 miles) at the equator and about 471 km (292 miles) at 45° latitude.

There are many types of map projections and each has its advantages and disadvantages. Some projections are well suited for constructing world maps. Other projections are designed for continents and countries. Some projections work well for areas with their longest extent in the east–west direction and others work better for areas with their longest extent in the north–south direction. Some projections are designed for a pleasing presentation and other projections are designed to support technical measurements of distances, directions, or areas.

An in-depth discussion of the hundreds of map projections available with ArcGIS is beyond the scope of this book. This chapter discusses some commonly used map projections and their typical use. For more details on map projections, consult the ArcGIS help system. A well known reference book for map projections is Map Projections, A Working Manual, *1987 by John Snyder.*

With a map projection, you can present the nearly spherical surface of the earth on a flat surface, such as a printed map or a computer display. While a globe is the most accurate scale model of the earth possible, globes are bulky and cannot be easily transported. Also, it's impractical to build a globe large enough to show cartographic details such as streets in a city. In contrast, maps can be made at any map scale and rolled, folded, or bound into an atlas and carried around.

Factors for selecting map projections

To select a map projection, you need to consider several factors pertinent to the purpose and area of the map you wish to build. In brief, these are some of the factors that will influence your selection of a map projection.

- *Map purpose.* Maps constructed for a special purpose should use a map projection that preserves important projection characteristics. For example, navigation maps usually use the Mercator projection because this projection has the special property of representing paths of constant bearing as straight lines (called rhumb lines). Thematic maps that symbolize an attribute of an area, such as population density, should use a equal-area projection.

- *Map extent.* The area covered by your map will affect your selection of map projection. Several dozen map projections have been designed specifically for maps of the world, such as the Winkel-Tripel and Robinson projections. Maps of continents often use conic projections such as the Albers Equal Area projection.

- *Map orientation.* Whether you are mapping an area with an extent longer in the east–west or north–south orientation will influence your selection of map projection. Generally, conic projections are used for areas with a greater east–west extent. Transverse cylindrical projections are commonly used for areas with a greater north–south extent.

- *Latitude range.* The latitude range of your map influences the selection of a map projection. Cylindrical projections work well for mapping equatorial regions. For mid-latitudes, the map projections most often used are based on cylindrical and conic projections. Polar regions are usually mapped with planar projections.

- *National coordinate systems.* If you are building general-purpose maps that cover local areas, such as cities, counties, states, or provinces, and especially if these maps are to be used by government agencies, then a national coordinate system is a good choice. Two examples are the State Plane Coordinate System in the United States and the Great Britain National Grid for the United Kingdom. For regions not covered by a national coordinate system, the Universal Transverse Mercator system (UTM) is commonly used for large-scale maps.

Guidelines for selecting map projections

World and hemispheric maps

Map projections for global maps are challenging because severe distortion is unavoidable. These are a few common choices.

General-purpose world maps

The Winkel-Tripel projection is often considered to be the best overall projection for world maps. This is the standard projection for National Geographic Society world maps.

Thematic maps

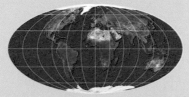

For thematic maps that display attributes based on areas of countries, it's important to use an equal area projection. The Mollweide projection is a common choice.

Hemispheric views

For hemispheric views, the orthographic projection simulates the view from space. It's a popular choice for locator maps that show the geographic extent for a detail map.

Continental and regional maps

Map projections for continents and regions are based on several factors: latitude range, map use, and orientation of map.

Consider the latitude range

Polar regions should be mapped with an azimuthal (planar) map projection.

A country or region in the mid-latitudes could be mapped with a conic map projection.

A country or region near the equator could be mapped with a cylindrical map projection.

Consider the map use

The azimuthal equidistant projection is used to show accurate airline travel distances from a center point.

The Mercator projection is used for navigational charts because straight lines represent lines of constant bearing.

The Albers equal area conic projection is widely used for thematic maps that require equal-area distribution.

Consider the map orientation

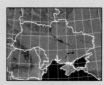

Maps of countries with greater east–west orientation such as Ukraine are best mapped with conic projections.

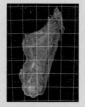

Maps of countries with greater north–south orientation such as Madagascar are best mapped with transverse cylindrical projections.

Large scale maps

Map projections for large scale maps of small regions and cities have low distortion of area and distance. Maps for these areas usually follow projected coordinate systems defined by regional or national mapping agencies.

This map of Munich, Germany, uses Germany zone 4, based on the Gauss-Krüger projection.

This map of Kyoto, Japan, uses zone 6 of the Japanese Geodetic Datum 2000.

This map of Sydney, Australia, is projected using Map Grid of Australian Zone 56.

Many map projections have been developed specifically for producing maps of the entire world. World maps are a special cartographic challenge because the entire world must be flattened onto a map sheet and this inevitably introduces severe distortions. Dozens of specialized map projections for world maps have been developed; a few of the important and commonly used projections are described here.

The Winkel-Tripel projection is a popular choice for world maps designed to minimize distortion of both distance and area. It's a compromise projection defined by averaging coordinates from the Equirectangular projection (cylindrical) and the Aitoff projection (modified azimuthal). The Winkel-Tripel projection is used by the National Geographic Society for world maps, replacing the Robinson projection which was previously used. It's considered the best overall projection for general-purpose world maps.

The Mollweide projection is an equal area projection frequently used for world maps showing altimetry and bathymetry. It's also used for maps of the sky showing stars and celestial bodies. This projection distorts shapes and preserves areas to form an elliptical outline with a 2:1 ratio between the horizontal and vertical axes.

The Sinusoidal projection (also known as Sanson-Flamsteed) distorts the shapes of the continents but does have the useful property of preserving equal areas and equally spaced horizontal parallels. The Sinusoidal projection is sometimes used for thematic world maps which symbolize an attribute proportional to area, such as population density of countries.

The Goode Homolosine projection is commonly used for school maps of the world and is an example of an interrupted map projection. This map projection divides the projections in the middle of the major oceans so that continents can be drawn with recognizable shapes. This projection maintains equal areas for continents of the world. A version of this projection interrupts the continents instead of oceans for producing maps of the world's oceans.

The Mercator projection is a frequent choice for navigational maps and sometimes used for world maps, but this projection introduces severe distortions for the Antarctic and Arctic regions. While Africa has about 13 times the land mass of Greenland, the Mercator projection makes them appear similar in size. If used for a world map, the map extent is usually limited to a latitude range such as 75° north to 75° south. While the Mercator projection is perhaps the best known projection and has a rectangular outline that conveniently fills a map sheet, it's not recommended for most world maps.

World map with Winkel-Tripel projection

World map with Mollweide projection

World map with Sinusoidal projection

World map with Goode Homolosine projection

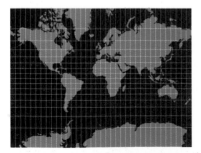

World map with Mercator projection

If you use conic projections with standard parallels (secant lines), this is a good rule of thumb to follow: Take the latitude range of your area of interest and divide into sixths. Define the first standard parallel at a latitude one-sixth of that range from the lowest latitude. Define the second standard parallel at a latitude one-sixth of that range below the highest latitude. For a latitude range from 30°N to 60°N, you would define standard parallels at 35°N and 55°N. Define the central meridian at the midpoint of the longitude range.

The common map projection used for the continents are one of the three major conic projections. The Albers Equal Area Conic projection is commonly used for middle latitudes, and as the name states, preserves equal areas. The Lambert Conformal Conic projection is widely used and is favored for aeronautical maps because it has the property of drawing great-circle routes as nearly straight lines. The Equidistant Conic projection is neither conformal or equal area, but is a compromise between the Lambert Conformal Conic and Albers Equal Area Conic projections.

Maps of the polar areas use planar projections, such as the Stereographic projection.

Conic projections of North America typically use standard parallels (secant lines) of 20°N and 40°N with a central meridian at 96°W. Maps of the 48 contiguous United States use standard parallels of 33°N and 45°N with a central meridian at 96°W, or if the Albers Equal Area Conic projection is used, standard parallels of 29.5°N and 45.5°N.

North America Albers Equal Area Conic

North America Equidistant Conic

North America Lambert Conformal Conic

Conic projections of South America can use standard parallels (secant lines) of 5°S and 42°S with a central meridian at 32°W.

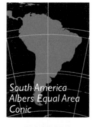
South America Albers Equal Area Conic

South America Equidistant Conic

South America Lambert Conformal Conic

Conic projections of Europe can use standard parallels (secant lines) of 43°N and 62°S with a central meridian at 10°E. Some maps of Europe use standard parallels of 35°N and 65°S with a central meridian at 10°E.

Europe Albers Equal Area Conic

Europe Equidistant Conic

Europe Lambert Conformal Conic

Conic projections of Africa can use standard parallels (secant lines) of 20° N and 23° S with a central meridian at 25° E. The Sinusoidal projection can also be used for Africa with a central meridian of 15° E.

Africa Albers Equal Area Conic

Africa Equidistant Conic

Africa Lambert Conformal Conic

Conic maps of Asia typically use standard parallels of 15°N and 65°N with a central meridian at 95°E. Some Lambert Conformal Conic projections use standard parallels of 30°N and 62°N with a central meridian at 105°E.

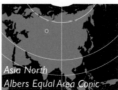
Asia North Albers Equal Area Conic

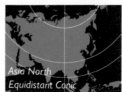
Asia North Equidistant Conic

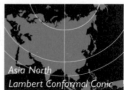
Asia North Lambert Conformal Conic

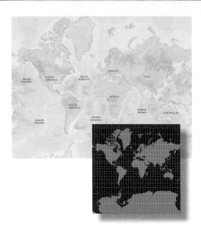

ArcGIS Online, Microsoft Bing Maps and Google Maps use a Mercator projection for two-dimensional map display.

The major online map services, such as ArcGIS Online, Microsoft Bing Maps and Google Maps, use the Mercator projection for displaying two-dimensional maps.

While the Mercator projection is not well suited for maps of the world because areas at high latitudes are severely distorted, the Mercator projection has become the standard for these mapping services because it works reasonably well at high zoom levels such as for cities or neighborhoods.

These are several reasons why the Mercator projection has been chosen for online mapping instead of other map projections:

- Online maps are stored with a continuous tiling system to support the seamless display of mapping data (imagery and street lines). This requires a single map projection for the world, instead of a series of zones such as in the UTM system.

- The Mercator projection preserves the orientation of the cardinal directions of north, south, east, and west. This is because the Mercator projection is cylindrical.

- The Mercator projection is conformal and preserves the shapes of features. This is important because objects such as city blocks and buildings must appear in their original form, whether square or rectangular.

An important limitation of how the Mercator projection is applied by online map services is that it is based on a sphere, instead of a spheroid. Recall that datums used for the world are based on a spheroid which has a polar axis about 1/3 of a percent shorter than the equatorial axis. The spherical Mercator projection (sometimes called Web Mercator) is used to optimize calculations and rapid display, but means that there is approximately a 0.33% scale distortion in the y-direction.

This distortion is not obvious on a computer screen, but would not be acceptable for technical map applications. Online map services are impressive and have revolutionized map use by the general public, but are not designed to replace maps for professional use produced by geographic information systems.

ArcGlobe and Google Earth presents a three-dimensional map display using the vertical perspective projection, which is a planar projection that simulates a view of the earth from a point in space. As you zoom in, the projection becomes an oblique perspective projection, similar to the vertical perspective, but with a mapping plane tilted with respect to the earth's surface.

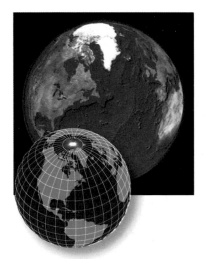

For three-dimensional displays, ArcGlobe and Google Earth uses the Vertical Perspective projection, a planar projection from the perspective of a point in space.

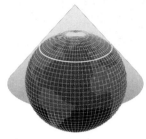

The simplest conic projection is tangent to the globe along a line of latitude called the standard parallel.

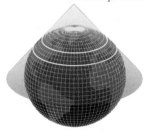

Somewhat more complex conic projections contact the global surface along two lines. These projections are called secant projections and are defined by two standard parallels.

Secant projections are widely used because they have the advantage of minimizing distortion caused by map projection over a larger area.

Conic projections are widely used, especially for middle latitudes and for mapping regions which have a greater west-east extent rather than a north–south extent.

The three most common conic map projections are the Equidistant Conic, Lambert Conformal Conic, and Albers Equal Area Conic.

Equidistant Conic projection

This conic projection can be based on one or two standard parallels. As its name implies, all circular parallels are spaced evenly along the meridians. This is true whether one or two parallels are used as the standards.

This conic projection is tangential if one standard parallel is specified and secant if two standard parallels are specified. Graticule lines are evenly spaced. Meridian spacing is equal, as is the space between each of the concentric arcs that describe the lines of latitude. The poles are represented as arcs rather than points.

Lambert Conformal Conic projection

This projection is good for middle latitudes. It is similar to the Albers Equal Area Conic projection except that Lambert Conformal Conic portrays shape more accurately than area. The State Plane Coordinate System uses this projection for all zones that have a greater east–west extent, such as the zone for the state of Tennessee.

Albers Equal Area Conic projection

This conic projection uses two standard parallels to reduce some of the distortion of a projection with one standard parallel. Although neither shape nor linear scale is truly correct, the distortion of these properties is minimized in the region between the standard parallels. This projection is best suited for land masses extending in an east-to-west orientation rather than those lying north to south.

The meridians are projected onto the conical surface, meeting at the apex, or point, of the cone. Parallel lines of latitude are projected onto the cone as rings.

The cone is then "cut" along any meridian to produce the final conic projection, which has straight converging lines for meridians and concentric circular arcs for parallels. The meridian opposite the cut line becomes the central meridian.

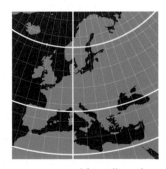

Conic projections are used for midlatitude zones that have an east–west orientation.

A cylindrical map projection projects the surface of the earth onto a cylinder. Like conic map projections, cylindrical map projections can be flattened without further distortion.

There are three aspects for cylindrical projections. The equatorial aspect has the cylinder of projection oriented along the equator and the Mercator projection is a well-known example of this type of cylindrical projection. The oblique aspect has the cylinder of projection oriented along any great circle on the earth and is used for specialized map projections for tilted areas such as the panhandle of Alaska. The transverse aspect has the cylinder of projection oriented along any meridian and is commonly used for projecting areas with a longer north—south extent, such as the U.S. State Plane zones for the state of Vermont. Several national systems and the Universal Transverse Mercator system are based on transverse cylindrical projections. Transverse cylindrical projections have the benefit of working well at all latitudes from the equator to the poles, provided the area of projection is not far from the meridian used for the projection.

Like conic projections, cylindrical projections can also have tangent or secant cases. Tangent cylindrical projection touch the earth along one line such as the equator or meridian. Secant cylindrical projections touch the earth along two secant lines and are commonly used because they have the benefit of minimizing projection distortion in the area between and near the two secant lines.

Three common cylindrical map projections are the Mercator, Transverse Mercator, and Equirectangular projections.

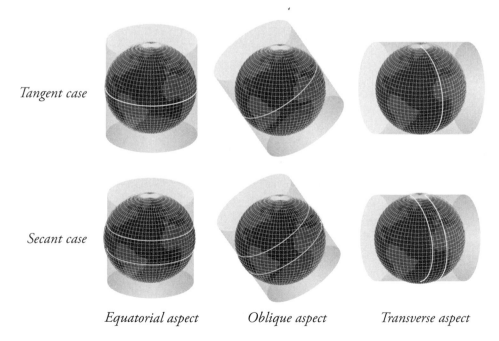

Tangent case

Secant case

Equatorial aspect *Oblique aspect* *Transverse aspect*

Mercator projection

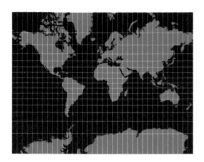

The Mercator projection was originally created to display accurate compass bearings for sea travel. An additional feature of this projection is that all local shapes are accurate and clearly defined. Meridians are parallel to each other and equally spaced. The lines of latitude are also parallel but become farther apart toward the poles. The poles cannot be shown.

This projection is conformal and displays true direction along straight lines. On a Mercator projection, rhumb lines, which are lines of constant bearing, are straight lines. While a rhumb line is not the same as a great-circle route, it is close enough for most sea navigation purposes and convenient for use with a compass.

The Mercator projection is one of the most common cylindrical projections. The equator is usually its line of tangency.

Transverse Mercator projection

The Transverse Mercator projection is one of the most important projections because it can be used for defining a series of zones, cylindrical and secant, about a meridian from pole to pole.

Many coordinate systems and map series are based on the Transverse Mercator projection; Universal Transverse Mercator (UTM), Gauss–Krüger, the U.S. State Plane Coordinate System, and others.

In the UTM coordinate system, the world is divided into 60 north and south zones six degrees wide. Each zone has a scale factor of 0.9996 and a false easting of 500,000 meters. Zones south of the equator have a false northing of 10,000,000 meters to ensure that all y values are positive. (The UTM system is discussed later as a topic in this chapter.)

The Gauss–Krüger coordinate system is similar to the UTM coordinate system. For example, Europe is divided into zones three degrees wide with the central meridian of zone 1 equal to 3° E. The parameters are the same as UTM except for the scale factor, which is equal to 1.000 rather than 0.9996. There are many variations of the Gauss–Krüger coordinate system in use.

The Transverse Mercator projection is conformal and does not maintain true directions. The central meridian can be placed in the center of the region of interest. This centering minimizes distortion of all properties in that region. This projection is best suited for north–south areas.

The Transverse Mercator projection is also known as Gauss–Krüger, and is similar to the Mercator projection, except that the cylinder is tangent along a meridian instead of the equator.

Equirectangular projection

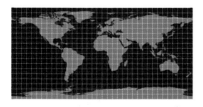

The Equirectangular projection is very simple to construct because it forms a grid of equal rectangles. Because of its simple calculations, its usage was more common in the past. In this projection, the polar regions are less distorted in scale and area than they are in the Mercator projection.

This simple cylindrical projection converts the globe into a Cartesian grid. Each rectangular grid cell has the same size, shape, and area. All the graticular intersections are 90 degrees. The central parallel may be any line, but the traditional Plate Carrée projection uses the equator. When the equator is used, the grid cells are perfect squares, but if any other parallel is used, the grids become rectangular. In this projection, the poles are represented as straight lines across the top and bottom of the grid.

The Equirectangular projection is also known as Simple Cylindrical, Equidistant Cylindrical, Rectangular, or Plate Carrée (if the standard parallel is the equator).

The orthographic projection views the globe from an infinite distance and gives the illusion of a three-dimensional globe. This projection is used more commonly for aesthetic display rather than technical presentation.

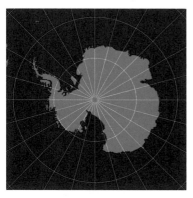

Antarctica mapped with the Lambert Azimuthal Equal Area projection.

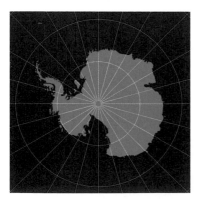

Antarctica mapped with the Azimuthal Equidistant projection.

Planar projections project map data onto a flat surface touching the globe. A Planar projection is also known as an Azimuthal or Zenithal projection.

This type of projection is usually tangent to the globe at one point but may be secant, also. The point of contact may be the North Pole, the South Pole, a point on the equator, or any point in between. This point specifies the aspect and is the focus of the projection. Possible aspects are polar, equatorial, and oblique.

Polar aspects are the simplest and most common form. Parallels of latitude are concentric circles centered on the pole and meridians are straight lines that intersect with their true angles of orientation at the pole. Directions from the focus are accurate.

Great circles passing through the focus are represented by straight lines; thus the shortest distance from the center to any other point on the map is a straight line. Patterns of area and shape distortion are circular about the focus. Planar projections are used most often to map polar regions.

Some planar projections view surface data from a specific point in space. The point of view determines how the spherical data is projected onto the flat surface. The perspective from which all locations are viewed varies between the different Azimuthal projections. The perspective point may be the center of the earth, a surface point directly opposite from the focus, or a point external to the globe, as if seen from a satellite or another planet.

Lambert Azimuthal Equal Area projection

The Lambert Azimuthal Equal Area projection preserves the area of individual polygons while simultaneously maintaining a true sense of direction from the center. The general pattern of distortion is radial. It is best suited for individual land masses that are symmetrically proportioned, either round or square.

Lambert Azimuthal Equal Area is projected from any point on the globe. This projection can accommodate all aspects: equatorial, polar, and oblique.

Azimuthal Equidistant projection

The most significant characteristic of the Azimuthal Equidistant projection is that both distance and direction are accurate from the central point. This projection can accommodate all aspects: equatorial, polar, and oblique.

As with other planar projections, the world is projected onto a flat surface from any point on the globe. Although all aspects are possible, the one used most commonly is the polar aspect, in which all meridians and parallels are divided equally to maintain the equidistant property. Oblique aspects centered on a city are also common.

Azimuthal projections are classified in part by the focus and, if applicable, by the perspective point. The Gnomonic projection views the surface data from the center of the earth, whereas the Stereographic projection views it from pole to pole. The Orthographic projection views the earth from an infinite point, as if from deep space. Note how the differences in perspective determine the amount of distortion toward the equator.

Orthographic projection

Stereographic projection

Gnonomic projection

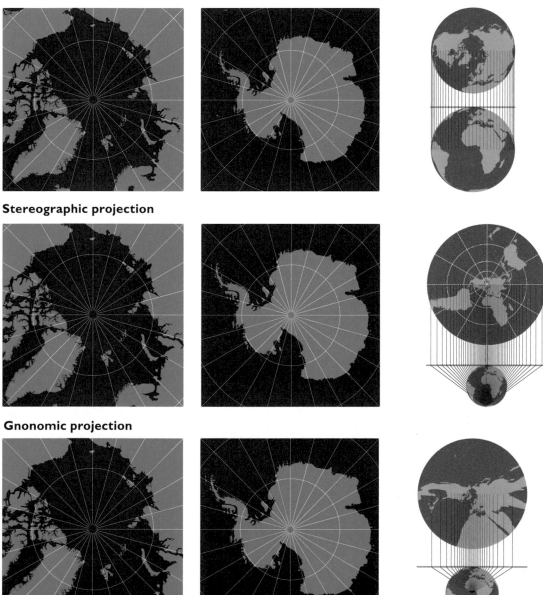

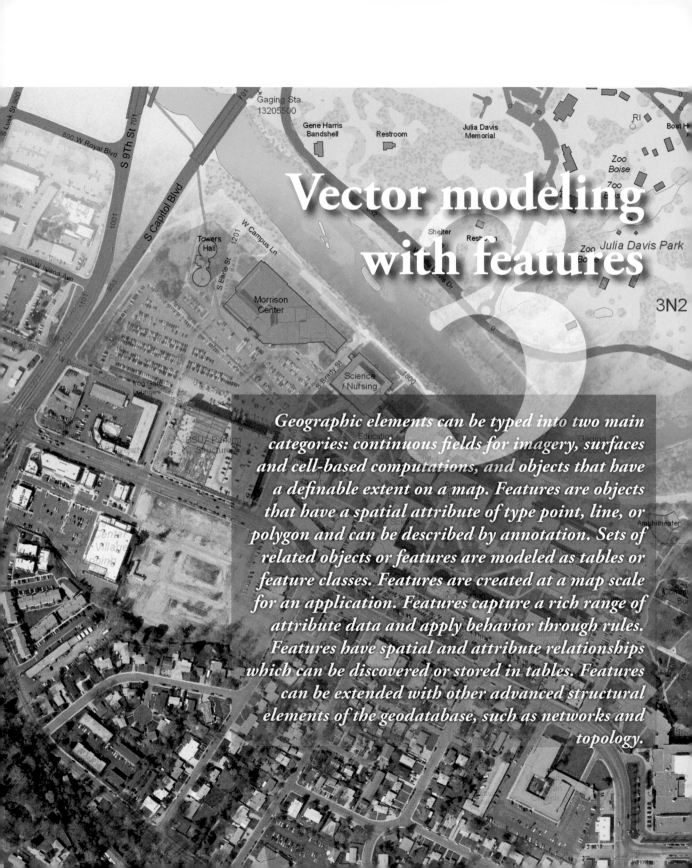

Vector modeling
with features

3N2

Geographic elements can be typed into two main categories: continuous fields for imagery, surfaces and cell-based computations, and objects that have a definable extent on a map. Features are objects that have a spatial attribute of type point, line, or polygon and can be described by annotation. Sets of related objects or features are modeled as tables or feature classes. Features are created at a map scale for an application. Features capture a rich range of attribute data and apply behavior through rules. Features have spatial and attribute relationships which can be discovered or stored in tables. Features can be extended with other advanced structural elements of the geodatabase, such as networks and topology.

A map is a scale model of a geographic area.

Topographic maps like this early USGS quadrangle map form a 'paper GIS'. By examining this map, you can perform many operations that you would like to do with a geodatabase. Spatial operations such as 'find the shortest route', 'what is the height of a hill', and 'what watershed does a stream drain' can be done by inspection and measurement of geographic features on this map.

Maps work because they clearly present the map layers of interest and follow cartographic conventions that everyone recognizes. Motorists easily navigate road maps, with highways, residential roads, and dirt roads, identifiable by their common line symbols. Contour lines effectively visualize slope angle by bunching up along steep mountain sides and spreading wide on the plains. The extent of a watershed can be found by deriving ridge lines around the stream that drains an area.

The spatial types of geographic features drawn on maps throughout history can sorted by dimension—points, lines, and areas—and are described by annotation on the map.

The map detail below is from a 1903 United Stated Geologic Survey (USGS) quadrangle map at 1:24,000 scale. The quadrangle map series, based on continuous grids of 15 minute and 7½ minute intervals of arc, was initiated by John Wesley Powell, the second director of the USGS, to provide a national map for developing the country's resources.

Topographic maps provide an accurate picture of the shape of the land with man-made features, such as roads, buildings, and survey points.

Since historic maps are drafted, areal features are drawn with fill patterns, elegantly rendered below in the tidal and bay water bodies. Map layers are distinguished by a standardized set of colors.

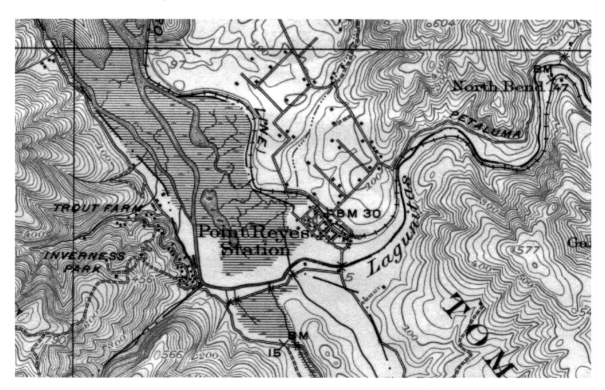

The map legend confirms how geographic features are organized. Maps represent discrete geographic entities with point, line, and area symbols.

Point symbols for small features and locations

Points on a map represent geographic entities too small to be depicted as lines or polygons, or locations at which a measurement is made.

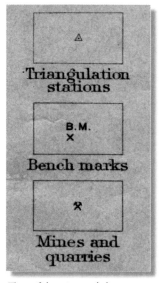

Three of the point symbols common on early USGS quadrangle maps.

Whether an entity is drawn as a point or other symbol depends on two factors: its greatest extent and the map scale of compilation. For example, small buildings are drawn with a small square point symbol, but larger buildings are drawn with a polygon of its outline.

Line symbols for long features and boundaries

Lines on a map represent geographic entities which are too narrow to depict as areas at the map scale of compilation.

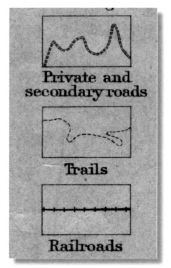

These line symbols for roads, trails, and railroads are still applied in today's topographic maps.

Lines are also used for computed lines such as survey traverses and graticules, or isolines of a given attribute, contour lines being an example.

Sets of connected lines form networks, such as roads and streams.

Area symbols for wide features and land divisions

Polygons on a map enclose areas that represent the shape and location of homogeneous areas, such as lakes, forest stands, and soil types.

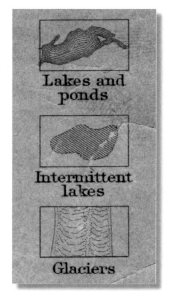

Early topographic maps used area symbols for natural features such as lakes; modern topographic maps also use area fills to mark regions such as urban areas.

Polygons are also the foundation for divisions of land, ranging from country boundaries to land parcels. A parcel fabric is a system in which every point within a jurisdiction is covered by exactly one polygon.

Feature information models then and now

This feature information model from 1903 and far earlier is fundamentally the same as today's feature information model in a geodatabase.

The three main feature types in a geodatabase are points, lines, and polygons.

When geographers compile maps, we make many decisions about how to represent features on a map. While geographic reality is infinitely complex, we build finite model features in a geodatabase, detailed enough to create maps and perform geoprocessing tasks, yet simple enough to keep data modeling and collection costs manageable.

Some of the considerations for whether and how geographic entities exist as features in a geodatabase are:

- Is a given feature of sufficient size or significance to be included on the map? Does it belong to a type classification of interest?

- Is a feature best represented by a point, line, or polygon shape?

- How accurately do features need to be positioned on the map?

Feature definition is guided by the map scale at which features are drawn. This principle is true whether map features are scanned from manuscript maps, derived from remote imagery, or collected with satellite positioning technologies.

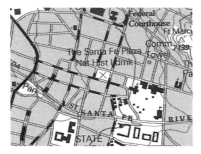

Features on a USGS quadrangle map at 1:24,000 scale.

Each set of features collected in a feature class is created for a map scale. This map scale determines which features are represented, whether they are represented with points, lines, or polygons, and the density of vertices that define shapes.

Shape types for features

While it may seem obvious for many features which shape type is correct, sometimes the answer is not clear.

Should a building be represented by a polygon or point? It depends on the map scale. If you are making a city map at large scale, buildings are frequently represented with polygons. If your map is at an intermediate scale, then most buildings might be represented as points with a few large buildings, such as hospitals and shopping malls, depicted with their polygonal outline. Small-scale maps might have groups of buildings aggregated into polygons representing urbanized areas.

The same geographic center on a USGS quadrangle map at 1:100,000 scale. A scale difference factor of about four drastically alters features representation and dimensional types.

Should a road be represented by a line? Usually, but for large-scale maps such as subdivision plans, roads are represented by polygons with boundaries at the curb lines. Even when a road is represented by lines, should the line represent the center line of a road or two parallel lines for each roadway in a divided highway? Again, it depends on the map scale.

Map scale also guides whether an entity is represented as a feature or as a row in a table. For example, a pole-mounted electric transformer is an important component of electric distribution circuits but rarely appears as a feature on maps of electric systems. The reason is that at most map scales, a transformer's position is indistinguishable from the pole's position. Therefore, only the pole is mapped as a feature, and associated equipment is represented in related tables for transformers and other devices.

In natural systems such as stream networks, a continuous range of objects exist. For example, how do you differentiate a large stream from a small river? No clear break point exists in nature for these features, so geographers define criterion for feature inclusion, such as a state (whether a body of water is perennial or intermittent), and defining minimum areas for lakes, minimum widths for rivers, and minimum lengths for stream segments.

To understand features, it's worth examining maps at several scales.

Map features at neighborhood scale

Neighborhood map scales in common use range from about 1:600 for civil engineering projects to perhaps 1:5,000 scale for maps focused on a neighborhood.

In this map, the river is drawn as a complex set of polygons, bordered by the river banks and including sandbar islands and oxbow lakes.

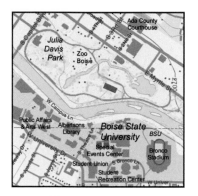

Detail from a planimetric map of Boise, Idaho, at 1:5,000 scale.

Roads are drawn as polygons, the road area being bounded by the curb lines. At this scale, the width of the road is directly visible and important roads are distinguished by their width, not a line symbol. Even pedestrian pathways are included on maps at this scale.

Interestingly, buildings are represented as both point features for houses and as polygon features for significant structures such as schools and courthouses.

Map features at city scale

Maps that span a sizeable city might range from about 1:12,000 to 1:50,000 scale. Features depicted at neighborhood scale undergo a transformation at typical city map scales.

Map section from Ada County, Idaho, at 1:50,000 scale.

Roads are drawn with center lines instead of areas and arterial roads are displayed through thicker and colored line symbols. All but minor roads are drawn.

Only important buildings such as hospitals are drawn with point symbols. Most buildings disappear and are replaced by land use areas, such as residential areas, parks, and industrial zones.

Most rivers are drawn with lines instead of polygons and details such as small streams and islands are lost. Point density of streams is simplified.

Map features at regional scale

Features on maps that cover a state or province undergo a great simplification in contrast to the larger-scale city and neighborhood maps.

Portion of statewide map of Idaho at 1:250,000 scale.

At typical scales from 1:100,000 to 1:1,000,000, most roads are dropped except for major roads and highways.

Only significant rivers and lakes appear. Information about land use is reduced to outlined areas of developed or incorporated jurisdictions.

A large and complex geographic entity such as an airport is reduced to a single point feature with annotation.

Clearly, the map scale of compilation and intended use is the starting point for thinking about how to define features. This sets the context for spatial type, detail, accuracy, and symbolization.

Geographic objects in the world exist within a rich context. They occupy a position, delineation, or area; have neighboring objects; may influence surrounding objects as a consumer or provider of a resource; have attributes that can be values, counts, categories, or descriptions; and might have a predictable response to an external stimulus.

Five qualities of features

Features, as represented in the geodatabase, have five prime qualities: they occupy a geometric shape in space, they have attributes stored in a table, they have rules to enhance data integrity, they have associations with other features and nonspatial objects, and the simple feature information model can be extended in a variety of ways for rich data modeling.

The subject of this chapter is the first four of these feature qualities: shape, attributes, behavior, and relationships. Subsequent chapters will treat the extensions to the feature information model: networks, locators, terrains, parcel fabrics, and more.

features have shapes

features are defined within a **spatial reference**

major shape types are **point, line,** and **polygon**

m-values on lines enables **linear referencing**

multipatch features and **z-values** model **3D objects**

features have attributes

tables contain a set of **rows** with **fields** to describe attributes

attribute values are validated through **domains**

subtypes allow fine control of feature behavior

features have behavior

attribute rules apply range and value domains

relationship rules specify the cardinality of relationships

topology rules define allowable spatial relationships

network rules set how network elements connect

features have relationships

spatial integrity of features is verified through **topologies**

joins and **relates** provide simple and dynamic data connections

relationship classes link features and nonspatial entities

features are extensible

quality mapping is done with **cartographic representations**

connecting lines can be organized into **networks**

a street network plus addressing rules create a **locator**

features with surface values form a **terrain dataset**

a network of land measurements forms a **parcel fabric**

versioning models feature transactions through time

Features have shapes

A feature class is a table in a geodatabase with a special field for shapes. There are five feature types for feature classes:

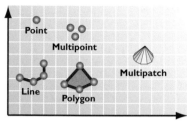

- Point features, for small features and locations where measurements are made
- Multipoint features, for efficient storage of many point measurements
- Line features, for simple or multipart linear features and boundaries
- Polygon features, for simple or multipart area features and divisions of land
- Multipatch features, for three-dimensional objects

Each feature shape has one or a set of x, y or x, y, z coordinates which is spatially referenced on the earth's surface through a coordinate system. Line features can also have measurement values for calibrating route systems.

Features have attributes

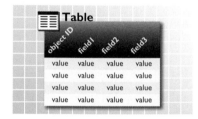

A feature class has a set of attributes for each feature in a table. Fields define feature attributes and can be numeric, textual, or descriptive. Some fields are predefined; most are user defined. Attributes can store rich content such as photographs.

To validate the accuracy of attribute data collection, each field of a feature class can have a domain, which is a numeric range or list of valid values. Fields can also be set with a default value when an object is created.

Features have behavior

Rules enforce data integrity within and among datasets. Rules apply to feature shapes, attributes, and relationships.

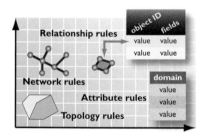

You use attribute rules to apply a domain to a field, relationship rules to constrain valid cardinality in relationships, topology rules for spatial relationships such as touches, contains, and overlaps, and network rules for validating which features can be connected.

Features have relationships

All geographic objects have some relationship to other objects. Features can also be related on-the-fly through joins on keys in tables. Joins and relates are a good way to dynamically associate features that you are working with on a map. Joins and relates are similar; joins associate information during an ArcMap session and relates preserve these associations over many ArcMap sessions in a map layer.

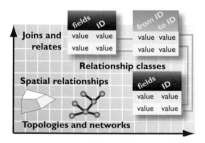

When necessary, you can define relationship classes among geographic objects in different feature classes and tables, such as the relationship between a house and its owner. Relationship classes help ensure data integrity. For example, the deletion or modification of one feature could delete or alter a related feature. Furthermore, since a relationship class is stored in the geodatabase, it is accessible to anyone who uses the geodatabase.

You can think of a GIS as an extension of database technology that stores, manages, and updates spatial information. Features are spatial objects and much of the functionality inside ArcGIS involves the display, query, and editing of features.

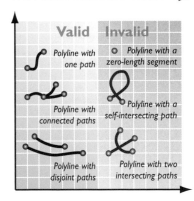

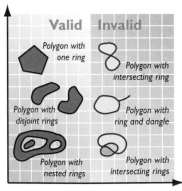

A feature class has a special field that represents the shape and location of features. This field is called shape and is of the field type geometry. All features in a feature class have the same type of geometry. A shape field of a feature class can be one of the following types of geometry: point, multipoint, polyline, or polygon.

Point geometry for point and multipoint features

A point feature has a point shape with a single x,y or x,y,z coordinate value. Sometimes points also have m-values when they are created as features from point events on a route (see Chapter 5, 'Linear referencing with routes').

A multipoint feature has a set of x,y or x,y,z coordinate values. There is no order implied in the set of coordinates in a multipoint shape. No two points in a multipoint can share the same x-, y-, z-, m-, and ID values. Multipoints with m-values could be created from event geoprocessing on a route.

Polyline geometry for line features

A line feature has a polyline shape consisting of one or more paths. A path is a connected collection of segments, each of which can be one of these types of parametric curves: line, circular arc, elliptical arc, or Bézier curve.

An optional z-value (commonly an elevation) or m-value (measurement distance) can be associated with a feature with polyline geometry. Line features with z-values can be added as features such as breaklines into terrain datasets. Line features with m-values are used as routes with linear referencing.

A polyline is a sequence of points, called vertices, connected pair-wise using segments collected into one or more paths. A path is a simply connected set of segments.

Paths in a polyline cannot be self-intersecting. Paths in a polyline cannot intersect other paths. Paths in a polyline can touch only at path endpoints.

Polygon geometry for polygon features

A feature with a polygon shape has one or more rings. A ring is a connected, closed, and nonintersecting set of segments. Each segment can be of type line, circular arc, elliptical arc, or Bézier curve.

A ring cannot intersect itself, but can intersect other rings in a polygon. Rings in a polygon can touch at any number of points.

An optional z-value (commonly an elevation) can be associated with a feature with polygon geometry. Polygons with constant z-values are commonly used to represent features such as lakes within a terrain dataset.

If rings in a polygon represent holes, they must be ordered in a direction opposite the boundary ring.

Components of feature shapes

It's worth noting that there are two types of geometry objects, those on which feature shapes are based (points, multipoints, polylines, and polygons), and those which combine to form feature shapes (segments, paths, and rings). In brief, this is the geometry model inside ArcGIS.

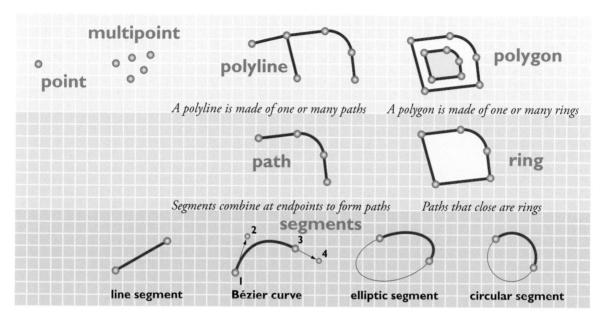

A polyline is made of one or many paths   *A polygon is made of one or many rings*

Segments combine at endpoints to form paths   *Paths that close are rings*

A segment is a function that describes a curve from one vertex to the next. There are four types of segments:

- A line is a straight segment between two points. Lines are used for straight constructions, such as a road segment or parcel boundary.

- A circular arc is part of a circle. It is commonly used for road curb lines at street intersections.

- An elliptic arc is part of an ellipse. It is used for the geometry of a transitional feature, such as a highway ramp.

- A Bézier curve is a parametric function defined by a set of third-order polynomials through four control points.

Polyline and polygon shapes can contain segments of all four types.

Tables and feature classes have types you select and properties that you set in building a geodatabase.

Tables

Tables in the geodatabase work in the same way as tables in any relational database management system. The only difference is that tables in the geodatabase contain an additional field called ObjectID, which is managed by ArcGIS to guarantee a unique identifier for every row in the geodatabase.

A table is a collection of rows, each with the same set of attributes, described by fields. Each row in a table represents a non-geographic entity, such as an owner. Attributes are defined with 12 field types which model a diverse range of information.

Feature classes

A feature class is simply a table with one special field; the shape field can have a feature type of point, line, polygon, multipoint, or multipatch. Each feature in a feature class has a geometry which describes its shape within a coordinate system. All features in a feature class have the same feature type.

Along with feature type, feature classes have geodatabase controlled fields for geometric properties such as length and area. Points, lines, and polygon have optional z- and m-values. While there are seven feature types, five geometry types —point, multipoint, polyline, poly-gon, and multipatch—are applied to features. Annotation and dimension feature classes are based on point and line shapes.

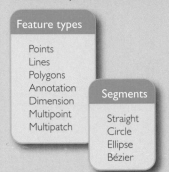

A table has a name, alias, and a collection of fields

A feature class is a table with a shape field of one of the feature types. It has geometry properties (z- and m-values), and a spatial reference that defines its coordinate system and resolution.

Field types

ObjectID
Short integer
Long integer
Float
Double
Text
Date
Blob
GUID/Global ID
Raster
Geometry
XML

Field properties

Field name
Alias
Allow NULL values
Precision
Scale
Default value
Domain
Length
Description
Managed by GDB
Geometry type

Feature types

Points
Lines
Polygons
Annotation
Dimension
Multipoint
Multipatch

Segments

Straight
Circle
Ellipse
Bézier

Lines and polygons are composed of segments

Rows in a table and features in a feature class can be associated in many ways

Use joins and relates for most associations between rows. Joins and relates model one-to-one and one-to-many associations.

Use relationship classes to persist a relationship between object in a geodatabase such as a feature class and a table. Relationship classes are stored as an object in the geodata-base and can model one-to-one, one-to-many, and many-to-many relationships. Relationship classes extend the relationship between rows and features in a geodatabase by storing attributes with the relationship and controlling how rows or features react when their associated rows or features are modified.

Use joins, relates, and relationship classes to manage associations between features and rows.

For associations between features and features, use joins, relates, or relation-ship classes if an association cannot be discovered or defined based on spatial relationships.

If a spatial relationship can be discovered or defined among features, use one of the operations or extended datasets below.

Use spatial joins to discover associa-tions among features using spatial relationships such as 'within', 'touches', or 'overlap'.

Use topologies to ensure that features maintain spatial integrity such as 'not intersecting' or 'not having gaps'.

Use geometric networks or network datasets to manage associations such as how features on a network are connected.

Use linear referencing to relate features to a route by measures.

Use parcel fabrics to manage a spatially integrated set of features such as land parcels covering an area.

Use terrain datasets to model surface relationships of point, line, and polygon features with z-values.

The geodatabase contains two fundamental datasets for modeling collections of objects: tables and feature classes. Collections of objects are modeled with tables if they are non-geographic and with feature classes if the entities have a definable shape on the earth.

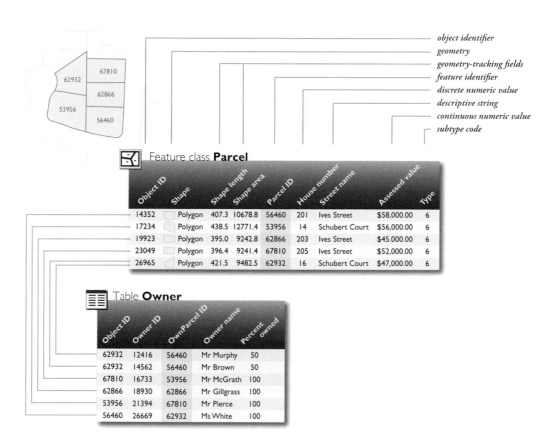

Land parcels and their ownership are neatly modeled by a feature class, a table, and a join. Each parcel is represented by a row with polygon shape in the Parcel feature class. Parcels are related to owners through a join associating values in the ParcelID field in the Parcel feature class with the OwnParcelID in the Owner table. Each owner is represented as a row in the Owner table. This simple data model supports associations such as a parcel with many owners.

These are some important ideas shown in this instance diagram:

- The simple feature information model comprising tables and feature classes with their associations can be used to effectively model real-world relationships.

- Fields in a table can handle a diverse set of data types—from simple numbers and text to global identifiers, blobs, and rasters.

- Joins and relates can model simple and complex relationships among features and objects.

Tables are the repository of objects and their attributes. A table stores attributes for objects that are reasonably similar to one another and have the same set of attributes.

These objects may be mobile (such as people or vehicles), measured values (such as environmental samples at a location over time), or they may be abstract entities (such as sales transactions).

Attributes express qualities of an object. These are some common types of attributes:

- An attribute can designate a coded value for a classification.

- An attribute can be descriptive text that characterizes a feature or gives it its name.

- An attribute can characterize a real numeric value that is measured or calculated, such as distance or flow.

- An attribute can specify a unique identifier that references a row in another table.

Table name and alias

Each table has a name and an alias. A table name is restricted to alphanumeric characters and underscores. Spaces, punctuation marks, and special characters are not allowed in a table name.

An alias is for an optional rich description of the table. Aliases can contain any character.

Rows and columns

A table is organized into rows and columns.

A row is the fundamental unit of information in a table and comprises a set of properties for an object. All the rows in a table must have the same set of property definitions.

A column represents all the attributes of one type. The value of a column for a given row is called an attribute. The definition of a column—its name and whether the column is formatted to store an object identifier, geometry, real numeric value, integer numeric value, or character string—is called a field.

Every field has a type such as number, text, or identifier, and this type is used to capture the various characteristics of an object. Generally, measured attributes are represented with number fields and descriptive attributes are represented with text.

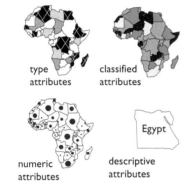

type attributes classified attributes

numeric attributes descriptive attributes

Egypt

On maps, feature attributes are displayed in distinct ways, depending on the type of attribute.

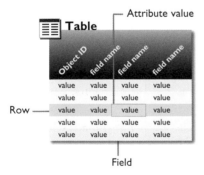

Attribute values for a table are arranged in a matrix of rows and fields.

Subtype

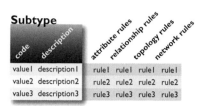

code	description	attribute rules	relationship rules	topology rules	network rules
value1	description1	rule1	rule1	rule1	rule1
value2	description2	rule2	rule2	rule2	rule2
value3	description3	rule3	rule3	rule3	rule3

Many rules of all types can be assigned to a subtype of a feature class.

✛ Feature class

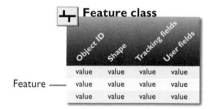

	Object ID	Shape	Tracking fields	User fields
	value	value	value	value
Feature	value	value	value	value
	value	value	value	value

↻ Relationship class

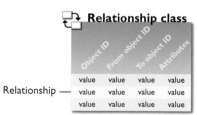

	Object ID	From object ID	To object ID	Attributes
	value	value	value	value
Relationship	value	value	value	value
	value	value	value	value

Rows in feature classes are features. Rows in relationship classes are relationships.

Subtypes

Tables can have a special field defined for subtypes. Subtypes are a logical division of the objects in a table for the purpose of exercising fine control over rules. Subtypes have an integer code value with a text description.

Subtypes are essential for data modeling because they let you minimize the number of tables and feature classes for a rich data model.

Object IDs and predefined fields

All tables in a geodatabase have a predefined field called Object ID. This identifier is a persistent and unique identifier for each row within the scope of a table. You should never use database software to modify Object ID values or other predefined fields.

Users often establish a separate object identifier which can be managed by a user application. These identifiers are used as keys for relating rows in pairs of tables and features classes.

Datasets based on tables, such as feature classes, have predefined fields for storing geometry properties such as distance and area.

Attribute indexes

You can create attribute indexes on fields to make query performance quicker. In ArcCatalog, you can create indexes on one or several attributes in a table and you can add and remove indexes at any time.

Only create those attribute indexes that you really need—performance diminishes when you define an excessive number of indexes.

Feature classes and relationship classes

In a geodatabase, datasets based on tables can store nonspatial objects, spatial objects, and relationships.

A table that stores spatial objects is a feature class. Simple feature classes have two predefined fields: an Object ID and a geometry field.

Annotation, dimension, and network-specific feature classes have additional predefined fields.

A table that stores relationships is a relationship class. It can have any number of custom fields to represent the attributes of the relationship.

Not all relationship classes are implemented as tables. If a relationship class is not attributed and does not have a cardinality of many-to-many, it is stored as a set of keys on the feature or object classes.

A feature class contains a set of thematically related features of the same spatial type.

A feature class has a special field called Shape to store the geometry of features.

All features are located with x and y coordinate values, with optional z-values for elevations and m-values for line measures.

Feature shapes are positioned on the world through a spatial reference.

A feature class has all the properties and behaviors of tables.

A feature class is a collection of features representing the same geographic elements, such as wells, parcels, or soil types. All the features in a feature class have the same spatial representation (for example, point, line, or polygon) and share a common set of descriptive attributes.

Individual features in a feature class can also share spatial relationships with other features. For example, adjacent polygons share boundaries according to well-defined integrity rules (such as that counties cannot overlap one another).

Linear features often participate in an interconnected network for analytical use (for example, a street network).

Name and alias

Each feature class in a geodatabase has a unique name. The name of a feature class cannot contain spaces, punctuation, or start with a number. You may optionally give the feature class an alias that acts as an alternate name. Unlike the feature class' true name, an alias may contain spaces, punctuation, and start with a number. If the feature class name contains abbreviations, an alias can expand that name into a readable form.

Feature types

Each feature class has a feature type, which specifies the geometry type stored in the Shape field. The most commonly used feature types are point, line, and polygon.

Line and polygon features have the flexibility to represent either single-part or multipart features. For example, one polygon feature can represent a complex object such as a chain of islands.

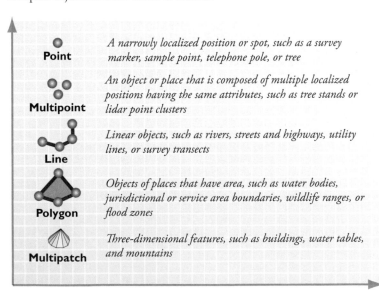

Point — *A narrowly localized position or spot, such as a survey marker, sample point, telephone pole, or tree*

Multipoint — *An object or place that is composed of multiple localized positions having the same attributes, such as tree stands or lidar point clusters*

Line — *Linear objects, such as rivers, streets and highways, utility lines, or survey transects*

Polygon — *Objects of places that have area, such as water bodies, jurisdictional or service area boundaries, wildlife ranges, or flood zones*

Multipatch — *Three-dimensional features, such as buildings, water tables, and mountains*

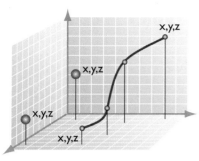

Feature coordinates can include x, y, and z coordinates.

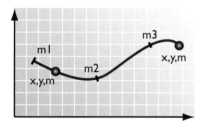

Dynamic segmentation is the process of computing the shape of route locations along calibrated linear features at runtime based on event tables for which distance measures are available.

Z- and m-values are not exclusive, vertices for measurements can be either (x,y,m) or (x,y,z,m).

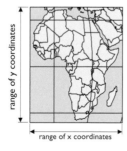

A spatial domain defines the maximum map extent in x and y coordinates.

Three dimensions with z-values

While all features are stored with x and y coordinate values, you can add z-values so that each point and vertex is located in three-dimensional space. Z-values are most commonly used to represent elevations, but they can represent other measurements such as annual rainfall or measures of air quality.

Z-values can be applied to features such as stream lines, ridge lines, or lakes. A ridge line is a profile along a surface. You can assign individual elevations at each point along a ridge line. A lake polygon would have identical z-values along the perimeter of the lake.

Multipoints with z-values are used for bulk loading many lidar points, a detailed x,y,z point cloud of the surface. Z-values prepare features with elevation data for input into a terrain dataset.

Linear referencing with m-values

Linear feature vertices can also include m-values to support linear referencing. Some GIS applications employ a linear measurement system used to interpolate distances along linear features, such as along roads, streams, and pipelines. A common example is a highway milepost measurement system used by departments of transportation for recording pavement conditions, speed limits, accident locations, and other incidents along highways.

A calibrated linear feature, or route, is a line feature that has m- (measure) values and an identifier. Route locations are organized into tables based on a common theme, either point or line event tables. For example, five event tables containing information on speed limits, year of resurfacing, present condition, signs, and accidents can reference a route feature class representing highways.

Spatial reference

Features in a feature class are tied to locations on earth through a coordinate system, which defines the map projection used and its parameters. All features in a feature class have the same spatial reference. A spatial reference combines a spatial domain, the allowable range of x and y values, with a coordinate system.

Spatial references are associated with individual stand-alone feature classes or sets of feature classes in a feature dataset. Spatial references, coordinate systems, and map projections are covered in Chapter 2, 'Coordinate systems and map projections'.

When you design a data model, you will be making a series of decisions to optimize your geodatabase for utility, performance, accuracy, and ease of maintenance.

A feature dataset is a container of feature classes and other datasets that enforce a common spatial reference. Feature datasets are used to spatially integrate related feature classes. Their primary purpose is for organizing related feature classes into a common dataset for building a topology, a network dataset, a terrain dataset, or a geometric network.

There are additional situations in which users apply feature datasets in their geodatabases:

- To organize data access based on database privileges. Sometimes, users organize data access privileges using feature datasets. All feature classes contained within a feature dataset have the same access privileges. For example, users might need to use more than one feature dataset to segment a series of related feature classes to account for differing access privileges between users. Each group has editing access to one of the feature datasets and its feature classes, but no edit access for the others.

- To organize feature classes for data sharing. In some data sharing situations, collaborating organizations might agree on a data sharing schema for sharing datasets with other users. In these situations, people might use feature datasets as folders to organize collections of simple feature classes for sharing with others.

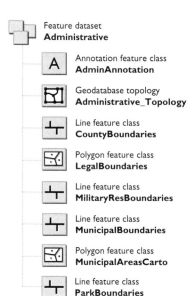

Feature dataset
Administrative

A — Annotation feature class
AdminAnnotation

Geodatabase topology
Administrative_Topology

Line feature class
CountyBoundaries

Polygon feature class
LegalBoundaries

Line feature class
MilitaryResBoundaries

Line feature class
MunicipalBoundaries

Polygon feature class
MunicipalAreasCarto

Line feature class
ParkBoundaries

A feature dataset is a container for feature classes that share the same spatial reference, along with relationship classes, networks, and geodatabase topologies.

Specifying the coordinate system

Another design factor in organizing feature classes into common feature datasets is the requirement to use a spatial reference. Thus, it's useful to define your coordinate system requirements for each feature class prior to organizing feature classes into common feature datasets.

When creating a new feature dataset, you must define its spatial reference. This includes its coordinate system—either geographic or a specific projection—as well as coordinate units and tolerances for x, y, z-, and m-values. All feature classes in the same feature dataset must share a common coordinate system, and x,y coordinates of their features should fall within a common spatial extent.

Many fields contain measurements. Understanding levels of measurement helps to select correct field types. For more on measurement frameworks, read Exploring Geographic Information Systems *by Nicholas Chrisman.*

Types of measurements are commonly classified into four levels: nominal, ordinal, interval, and ratio, with progressively more functional sets of operations. Operations on measurements range from equality and inequality to mathematical functions. These levels of measurement were formalized by Stanley Smith Stevens in his landmark 1946 paper, 'On the theory of scales of measurement'.

Classified data

Nominal data values are used to classify, identify, and categorize data. The value is arbitrary and represents a quality, not a quantity. Values are considered labels with no comparative meaning. The only valid comparisons among nominal data values are equality and inequality.

Maps that present nominal data values include soil maps, land use maps, parcel code maps, and countries of the world. Field types used for nominal data are short integer, long integer, and text.

The colors and symbols of geologic units have no comparative value, but define typed units.

Ranked data

Ordinal data values determine the rank of an entity versus other entities. As well as equality and inequality comparisons, ordinal data values allow greater-than and less-than comparisons, but no mathematical functions such as addition or multiplication.

Maps that display ordinal data values include street maps symbolized by road type and maps of marketing territories shaded by sales ranking. Field types preferred for ordinal data are short or long integer, but in practice, text is used as well.

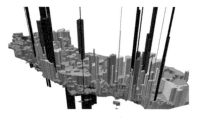

The hierarchy of road types in a city is displayed by color and line symbols.

Relative data

Interval data values represent a measurement on a calibrated scale, but with an arbitrary zero point, like the Fahrenheit temperature scale. Interval data values support addition and subtraction. Multiplication and division are performed on differences of values. Because the zero point is arbitrary, negative values can be used.

Maps that depict interval data values include temperature isoline maps and contour elevation maps. Appropriate field types are float, double, and date, with values sometimes rounded to short and long integers.

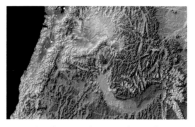

The physiographic map shows elevations against sea level, but not earth's center.

Absolute data

Ratio data values represent a measure on a scale with a fixed and meaningful zero point. Length, area, and the Kelvin temperature scale are examples of ratio data. Ratio data directly supports multiplication and divisions, and negative values are usually not allowed.

Maps that visualize ratio data values include weather maps with barometric pressure and rainfall, and chemical analyte concentration maps. The best field types for ratio data are float and double, but sometimes values are rounded to short and long integers.

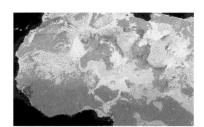

Projected growth from census data shown as extruded areas on a regional perspective view.

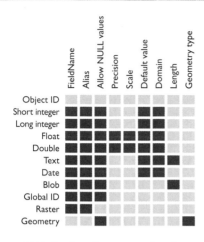

	FieldName	Alias	Allow NULL values	Precision	Scale	Default value	Domain	Length	Geometry type
Object ID									
Short integer	■	■	■			■	■		
Long integer	■	■	■			■	■		
Float	■	■	■	■	■	■	■		
Double	■	■	■	■	■	■	■		
Text	■	■	■			■	■	■	
Date	■	■	■			■	■		
Blob	■	■	■					■	
Global ID	■	■							
Raster	■	■	■						
Geometry			■						■

Fields describe a diverse range of attributes, from text and numbers to blobs, rasters, and geometries.

Field properties

In brief, these are the properties of fields applied to field types as shown in the diagram above:

- *FieldName* and *Alias* identify the field.

- *Allow null values* sets whether values must be entered.

- *Precision* and *Scale* set significant digits of numbers.

- *Default value* sets a value when a row is created.

- *Domain* identifies a list of valid values for this field.

- *Length* adjusts the field size for text and blobs.

- *Description* and *Managed by GDB* are raster properties.

- *Geometry type* is the field defining feature shapes.

Fields describe the attributes in a column.

Each field has a name and an alias. The field name must be unique within a table and is restricted to alphanumeric characters and underscores. Spaces, punctuation marks, and special characters are not allowed in a field name. Field aliases are a more descriptive name for a field, especially useful for names made of abbreviations.

Fields have additional properties, depending on the field type. For example, short integer, long integer, float, double, text, and date have default values and domains associated.

Discrete number values with short and long integers

Some attributes represent numeric values that are counted, not continuous. They are usually positive integer values, but can be negative in some cases. Short and long integers are designed for attributes which are counts, codes, and types, or any measured value which is rounded to an integer.

short integer 2 bytes	■□□□□□□□□□□□□□□□ *A short value contains 1 sign bit and 15 binary bits.*	range is -32,768 to 32,767
long integer 4 bytes	■□□□□□□□□□□□□□□□□□□□□□□□□□□□□□□□ *An integer value contains 1 sign bit and 31 binary bits.*	range is -2,147,483,648 to 2,147,483,647

If the range of anticipated integer values is well within a range of plus and minus 32 thousand, select short integer for the field type. Half the byte storage is required for short integers in contrast to long integers.

Short and long integer fields have properties of name, alias, allow null value, default value, and domain. For integers, a null value is not zero but an indeterminate and unassigned value.

A default value can be assigned to a short or long field when a row or feature is created. An integer field used for counting could have zero value for a default.

A domain can be specified for an integer field. Two types of domains are possible: coded-value domains and range domains. A coded-value domain for integers is a set of valid integer codes. A range domain for integers sets the minimum and maximum allowed values. Domains are discussed further, later in this chapter.

Continuous number values with floats and longs

A field can represent real numeric values, such as 2.7. Continuous numbers are used for data measured or calculated, such as distance or flow. Attributes using floats and doubles include length, route measure, and width.

float 4 bytes	A float value contains 1 sign bit, 7 exponent bits, and 24 mantissa bits.	7 significant digits, approximately -3.4 x 10^{38} to 1.2 x 10^{38}
double 8 bytes	A double value contains 1 sign bit, 7 exponent bits, and 56 mantissa bits.	16 significant digits, approximately -2.2 x 10^{308} to 1.8 x 10^{308}

Continuous real-valued numbers can be assigned to float or double field types. Float and double fields have properties of name, alias, allow null value, precision, scale, default value, and domain.

For floats and doubles, a null value is not 0.0 but an indeterminate value, which indicates missing or unknown data. If nulls are allowable, the allow null value property must be set in a new table before any rows are populated.

The field properties of precision and scale let you control the range of real values entered and the number format with which they are displayed. For example, if you wanted to constrain width values from 0.0 to 9999.99, then you would use a precision of 6 (the total number of significant digits) and a scale of 2 (to set the number of digits to the right of the decimal place).

For reporting many measurements, a small scale value is best because additional digits clutter a value. For example, temperature values are rarely reported with scale values greater than one (such as 98.6° Fahrenheit).

Default values are applied to an attribute when a new row is created in a table. For example, a concentration attribute could have a default value of 0.0 when a new measurement is made.

Precision and scale constrain the number of significant digits, but sometimes you need greater control of values. Float and long fields can be assigned a range domain, which constrains the minimum and maximum values for float and double fields. For example, elevation values on contour lines could be restricted from zero (sea level) to 9,000 meters.

Precision = 6

Scale = 2

4 2 1 . 6 6
1 6 . 9
7 8 3 2 . 0

For a precision of 6 and a scale of 2, these are valid numbers to enter. Note that the decimal place is not counted in the precision value. If you enter invalid numbers such as 63256 and 34.634, you will receive an error message.

Note that not all types of geodatabases support precision and scale properties.

Range domains let you restrict valid float or double values to precise minimum and maximum values.

Descriptions with text

If an attribute represents a finite list of text values, such as 'Fir', 'Maple', 'Pine', and 'Cedar' for types of trees, consider an integer field with a coded-value domain instead of a text field. When you associate a coded-value field with an integer field, the description will appear instead of the number code, geodatabase storage space is reduced, and the data is easier to maintain and validate.

An attribute can be a descriptive string that characterizes a feature or gives its name. Examples of text attributes are street names, soil types, descriptive notes, and owner names.

A text field contains a set of alphanumeric symbols.

To handle all the major languages of the world, text fields use the Unicode character handling system for storage and interchange of textual data. The Unicode system specifies a unique code for each character used in all major languages. This allows text to be converted among languages while retaining the correct set of glyphs (symbols representing characters and punctuation).

The properties of text fields are field name, alias, length, allow null values, default value, and domain.

The length of a text field is the number of characters allowed for the maximum text string anticipated. For place names, a length of 40 may be adequate, while for field notes, 80 or 240 characters may be desired for description.

Text field values can be null. A null value is indeterminate, and not the same as a blank text string. If null values are to be allowed, this property must be set when a table is created.

Text field values can also have a default value set. For example, a street feature class might have 'Road' for a default road suffix.

Time values with dates

MM/dd/yyyy

M/d/yy

M/d/yyyy

MM/dd/yy

yy/MM/dd

yyyy-MM-dd

dd-MMM-yy

These are some of the short date formats you can choose from.

Important events on objects can have a date value assigned that records a time. This is the temporal dimension of your GIS. When a road's pavement is replaced or maintained, a road inventory database can mark the time of that event with a date value.

The date field type can store dates, times, or dates and times. The default format in which the information is presented is mm/dd/yyyy hh:mm:ss with a specification of AM or PM. Date settings in the Microsoft Windows system determine the date format as reported in ArcGIS applications. For example, you can have year-month-date, month-date-year, or day-month-year.

The properties of a date field are field name, alias, allow null values, default value, and domain.

{3F2504E0-4F89-11D3-9A0C-0305E82C3301}

Sample GUID value with curly brackets and hexadecimal values.

Global IDs are used to track features in one-way and two-way geodatabase replication. Global IDs are also used by developers in relationships or any application requiring globally unique identifiers.

A photograph of a house can be stored with a building feature as a raster.

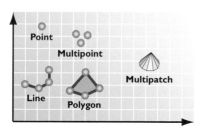

The shape field of a feature class can be set to one of seven feature types.

Unique identifiers with Object IDs

Every table and derived dataset in a geodatabase has an object ID field. The geodatabase manages these IDs and guarantees a unique identifier for every row in a geodatabase.

The field is named ObjectID and has an alias of OBJECTID.

Null values are not allowed for Object IDs. Object IDs are assigned by the geodatabase when a row or feature is created and this value never changes for the lifetime of that row or feature within the context of its table or feature class.

Global identifiers and GUIDs

Global ID and GUID data types store registry style strings consisting of 36 characters enclosed in curly brackets. These strings uniquely identify a feature or table row within a geodatabase and across geodatabases.

Global identifiers have a name and alias. Global identifiers can be set to allow null values.

Imagery with rasters

A feature can have a raster field, which is usually used for photographs of features, scanned blueprints, or document imaging. The properties of a raster field are field name, alias, description, and managed by the geodatabase, which sets whether rasters are directly managed in the geodatabase or stored in the computer's file system.

Feature shapes with geometries

Feature classes contain a special field called Shape. Depending on the feature class type, one of the dimensional geometry types is represented in this field. Annotation and dimensions are special features that describe other features.

Feature shapes and their specifications are discussed later in this chapter.

Every user's goal when adding or editing objects and features in a GIS database is to eliminate or minimize data entry error. For many data modelers, this is one of the most important aspects of designing a geodatabase.

Geodatabase feature classes store objects of the same type, that is, objects that have the same feature type (point, line, or polygon) and the same set of attributes. Although all objects in a feature class must have the same feature type and attributes, not all objects need to share the same behavior.

Subtypes

Subtypes are a powerful tool for data modeling because they let you control every configurable behavior in your geodatabase: attribute rules, relationship rules, network rules, and topology rules, without creating more tables. The result is a coarse-grained model, good for database performance, with finely discriminated object behavior.

A subtype is a special attribute that lets you assign distinct simple behavior for different classifications of your objects or features. All subtypes of a class share the same set of attributes.

The motivation for defining subtypes in tables and feature classes is to introduce a lightweight subdivision that adds these capabilities:

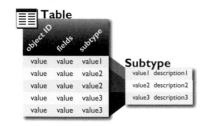

- You can name subtypes to describe each member of a classification of your objects.

- You can define distinct attribute domains for each field in a subtype.

- You can define distinct default values for each field in a subtype.

- You can prescribe the types of relationships that are possible between the objects in a subtype and objects in another subtype in the same or different object class.

- You can set different topological constraints on how feature interact.

A table or feature class does not have to contain subtypes. If none are defined, you can still set attribute domains, default values, and rules—but on the table or feature class as a whole instead of on a subtype.

Subtypes are classifications within a feature class or table. They allow you to logically group features based on a unique characteristic or behavior of the data.

Most importantly, you can apply feature and object behavior through rules at the subtype level. Subtypes are essential to good design because they help reduce the number of feature classes and improve performance.

Well-designed data models contain a manageable set of feature classes along with the means to finely control feature behavior. Subtypes are one way to limit the number of feature classes required in your data model, while giving you a powerful set of rules to ensure attribute, relationship, network, and topological integrity. When choosing between defining a set of similar feature classes or one feature class with the same number of subtypes, you should choose subtypes unless the set of attributes or geometry type is different.

Multiple feature classes or subtypes?

Tables are the cornerstone of relational databases, but an inevitable tension arises in data modeling—when is it better to split a set of features into distinct feature classes or lump them into one feature class?

Part of the answer comes from looking at the sets of attributes of similar features. If two features share 15 out of 17 attributes, then it is reasonable to group them together in the same feature class; you can tolerate a few disused attributes for certain features.

An overriding factor, however, is the geometry type; for vector features, you will typically choose point, line, or polygon for each feature. This forces distinct feature classes, since you cannot group features with different geometry types in the same feature class.

But other than these two situations—more than a few disused attributes and distinct geometry types—you should generally seek ways to use subtypes before creating additional feature classes.

	Subtypes in a feature class	Multiple feature classes
Collections	Subtypes define collections of features within a single feature class.	Feature classes store collections of features in the geodatabase.
Behavior	You can control behavior at the subtype level.	You can also control behavior by feature class.
Benefit	Subtypes reduce the number of feature classes.	Multiple feature classes give full modeling flexibility.
Rules	Subtypes define attribute domains, default attribute values, split–merge policies, connectivity rules, relationship rules, network rules, and topology rules.	Feature classes define attribute domains, default attribute values, split–merge policies, connectivity rules, relationship rules, network rules, and topology rules.
Attributes	Features in subtypes must have the same attributes.	Separate feature classes can have varied attributes.
Geometry types	Features in subtypes must share the geometry type (point, line, polygon).	Feature classes can have distinct geometry types.
Pros	Subtypes provide flexibility for applying rules to collections of features without requiring a separate feature class. Fewer classes result in fewer database queries for drawing and editing.	Multiple feature classes allow different attributes for collections of features. They also allow differing participation in topologies, networks, and relationship classes.
Cons	Subtypes cannot be used when a set of features has different attributes or spatial types.	Additional feature classes increase the number of queries required for operations such as drawing, editing, and topology validation.
Recommendations	Use subtypes liberally to reduce the number of feature classes in your data model.	Use multiple feature classes when you need more flexibility for defining attributes or geometry types.

A geodatabase provides a framework within which features can have behavior such as subtypes, default values, attribute domains, validation rules, and structured relationships to other objects. This behavior enables you to more accurately model the world and maintain referential integrity between objects in the geodatabase.

The behavior that you can control through subtypes on a table or feature class fall into four broad categories: attribute rules, connectivity rules, relationship rules, and topology rules.

Attribute rules

Attribute domains are rules that describe the legal values of a field type, providing a method for enforcing data integrity. Attribute domains are used to constrain the values allowed in any particular attribute for a table or feature class. A domain is a declaration of acceptable attribute values. Whenever a domain is associated with an attribute field, only the values within that domain are valid for the field. On editing, the field can accept a value that is not in that domain—validation will identify and let you correct rows with invalid values.

Attribute domains are discussed in more detail in the following topic.

Using domains helps ensure data integrity by limiting the choice of values for a particular field. Attribute rules are not applied during bulk loading; validation is done within ArcGIS. The geodatabase can be considered to be optimistic regarding data integrity during loading; ArcGIS lets you discover and fix invalid values.

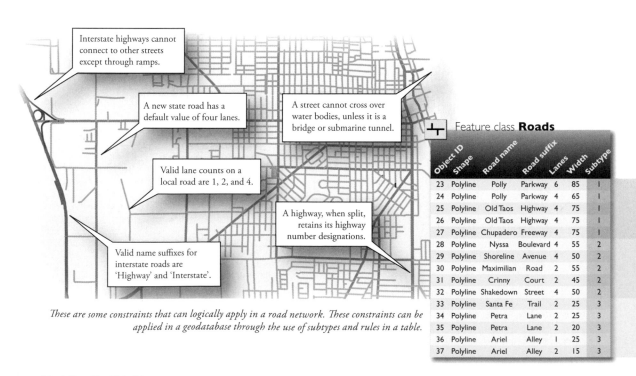

Interstate highways cannot connect to other streets except through ramps.

A new state road has a default value of four lanes.

A street cannot cross over water bodies, unless it is a bridge or submarine tunnel.

Valid lane counts on a local road are 1, 2, and 4.

A highway, when split, retains its highway number designations.

Valid name suffixes for interstate roads are 'Highway' and 'Interstate'.

Feature class **Roads**

Object ID	Shape	Road name	Road suffix	Lanes	Width	Subtype
23	Polyline	Polly	Parkway	6	85	1
24	Polyline	Polly	Parkway	4	65	1
25	Polyline	Old Taos	Highway	4	75	1
26	Polyline	Old Taos	Highway	4	75	1
27	Polyline	Chupadero	Freeway	4	75	1
28	Polyline	Nyssa	Boulevard	4	55	2
29	Polyline	Shoreline	Avenue	4	50	2
30	Polyline	Maximilian	Road	2	55	2
31	Polyline	Crinny	Court	2	45	2
32	Polyline	Shakedown	Street	4	50	2
33	Polyline	Santa Fe	Trail	2	25	3
34	Polyline	Petra	Lane	2	25	3
35	Polyline	Petra	Lane	2	20	3
36	Polyline	Ariel	Alley	1	25	3
37	Polyline	Ariel	Alley	2	15	3

These are some constraints that can logically apply in a road network. These constraints can be applied in a geodatabase through the use of subtypes and rules in a table.

Connectivity rules

Connectivity rules apply to geometric network features and are discussed further in Chapter 4, 'Linear modeling with networks'.

A connectivity rule constrains the type of network features that may be connected to one another and the number of features of any particular type that can be connected to features of another type.

For example, an electric line with phase ABC may be connected to a downstream line with phase AC.

The types of connectivity rules are edge–junction rule and edge–edge rule.

Relationship rules

Relationship rules are discussed further in the topic on relationship classes.

A relationship rule constrains the cardinality of a relationship between an origin class and destination class. The four basic cardinalities are one-to-one, one-to-many, many-to-one, and many-to-many.

With a relationship rule, you can create specialized cardinalities, such as a state has exactly two senators; a parcel of land can have no, one, or two owners; or a pole can have no, one, two, or three transformers mounted on it.

Topology rules

Topology rules are treated in detail later in this chapter.

Topology rules are used to ensure data quality and allow your geodatabase to more realistically represent geographic features. Unlike other feature behavior, topology rules are managed through a topology, not through individual feature classes.

	Default values	Attribute domains	Split / merge policy	Connectivity rules	Relationship rules	Topology rules
Highway	A new highway is given a default value of four lanes	Valid road suffixes are Highway, Parkway and Freeway	A split highway maintains all highway designations	A freeway can only connect to a ramp	A highway can be associated with two highway routes	A highway cannot be coincident with a river
Arterial	An arterial road has a default width value of 55 feet	Valid road widths are between 40 and 60 feet	A merged arterial road takes the weighted average for width	A four-lane arterial can only connect to another four-lane road	An arterial segment can be related with one highway route	An arterial road must not cross over a local road
Local	A local road has a default value of two lanes	Valid road widths are 15, 20, and 25 feet	A split local road retains the same number of lanes	A local road cannot directly connect to a freeway.	A private road can have up to two owners	A local road cannot cross a railroad

Attribute rules constrain the values of attributes.

A simple attribute rule is the default value for a field in a table or feature class. This lets you set a predefined default value, such as zero, instead of a null value.

Another attribute rule is setting an attribute domain to a field. Attribute domains describe the legal values of a field type, providing a method for enforcing data integrity.

An attribute domain is a declaration of acceptable attribute values. Whenever a domain is associated with an attribute field, only the values within that domain are valid for the field.

Attribute domains are not enforced during data loading, but validation commands discover and let you correct attribute values.

Attribute domains are stored at the geodatabase level, allowing them to be shared across feature classes, tables, and subtypes in a geodatabase.

When editing features and objects in ArcMap, you can enter features with invalid values, but you can validate your work at any time. Invalid attribute values are highlighted for editing.

Types of domains

There are two types of domains: coded-value and range.

Coded-value domains—A coded-value domain can apply to any type of attribute that supports domains—text, numeric, date, and so on. Coded-value domains specify a valid set of values for an attribute.

For example, water mains may be buried under different types of surfaces as signified by a GroundSurfaceType attribute field: pavement, gravel, sand, or none (for exposed water mains).

The coded-value domain includes both the actual value that is stored in the database (for example, 1 for pavement) and a more user-friendly description of what that value actually means.

Range domains—A range domain specifies a valid range of values for a numeric attribute. When creating a range domain, you enter a minimum and maximum valid value. A range domain can be applied to short-integer, long-integer, float, double, and date attribute types.

For example, in a feature class for water mains, you could have subtypes for transmission, distribution, and bypass water mains. Distribution water mains can have a pressure between 50 and 75 psi. For a distribution water main object to be valid, its pressure value must be entered as some value between 50 and 75 psi.

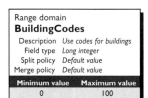

Range domain
BuildingCodes

Description	Use codes for buildings
Field type	Long integer
Split policy	Default value
Merge policy	Default value

Minimum value	Maximum value
0	100

Coded value domain
ClassCodes

Description	Class codes for streets
Field type	String
Split policy	Default value
Merge policy	Default value

Code	Description
A	Alley
C	Collector
E	Expressway
M	Major Road
S	Local Street
U	Unknown
P	Path

These are sample range and coded-value domains taken from a published address data model.

Properties of attribute domains

An attribute domain has a name, description, field type, split policy and a merge policy.

The field types for which an attribute domain can be set are short integers, long integers, floats, doubles, text, and date. A coded-value domain can be specified with all these types and a range domain can be specified with all these types except text.

Splitting and merging features

When features are split or merged, you can control what happens to numeric, date, and text fields.

For range domains, when features are split, you can choose to either duplicate an attribute value, apply the default value for that field, or calculate a geometric ratio based on proportional length or area measurements. When features with a range domain are merged, you can assign the new feature with a default value, sum the two values, or set a weighted average of the two values.

For coded-value domains, when features are split, you can duplicate the attribute value or set a default value. When features with a coded-value domain are merged, you can set a default value.

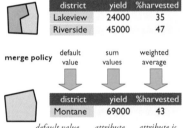

split policy duplicate / default value / geometry ratio

owner	land zoning	value
G Gould	R-4	25000

owner	land zoning	value
G Gould	R-1	14000
G Gould	R-1	11000

attribute of original feature is duplicated in split features / *default value is applied to split features* / *attribute is subdivided by ratio of split area or length*

district	yield	%harvested
Lakeview	24000	35
Riverside	45000	47

merge policy default value / sum values / weighted average

district	yield	%harvested
Montane	69000	43

default value for attribute is applied to the merged feature / *attribute is summed* / *attribute is weighted average of two attributes*

Steps to creating an attribute domain

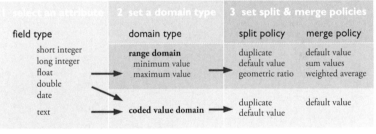

1 select an attribute	2 set a domain type	3 set split & merge policies	
field type	domain type	split policy	merge policy
short integer long integer float double date	**range domain** minimum value maximum value	duplicate default value geometric ratio	default value sum values weighted average
text	**coded value domain**	duplicate default value	default value

These are the field types on which attribute domains can be defined.

Text can only have a coded-value domain. All other attribute types can have a coded-value or range domain.

Based on the domain type, these are the valid split and merge policies available.

Point, line, and polygon features are annotated on this map detail.

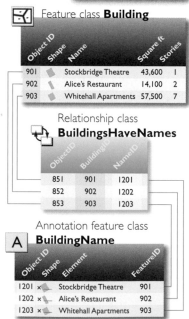

Three buildings have three feature-linked annotations.

An annotation is a type of feature that provides a textual description of a place or feature.

Annotation in the geodatabase is stored in annotation feature classes. As with other feature classes, all features in an annotation feature class have a geographic location and attributes and can either be inside a feature dataset or a stand-alone feature class.

The annotation feature class includes properties such as the field from which text labels are derived, the type of symbol, and other attributes. Each text annotation feature has symbology including font, size, color, and any other text symbol property. Annotation is typically text, but it can also include graphic shapes—for example, boxes or arrows—that require other types of symbology.

Standard and feature-linked annotation

There are two kinds of annotation in the geodatabase: standard and feature linked. Standard annotation is not formally associated with features in the geodatabase. An example of standard annotation is the text on a map for a mountain range. No specific feature represents the mountain range, but it is an area you want to mark.

Feature-linked annotation is associated with a specific feature in another feature class in the geodatabase. The text in feature-linked annotation reflects the value of a field or fields from the feature to which it's linked. For example, the water transmission mains in a water network can be annotated with their names, which are stored in a field in the transmission mains feature class.

Annotation classes

Standard and feature-linked geodatabase annotation feature classes contain one or more annotation classes. Each annotation class contains properties that determine how a subset of annotation in the feature class displays.

For both standard annotation and feature-linked annotation, these properties are:

- Default symbology applied when creating new annotation

- A visible scale range

For feature-linked annotation, the following properties are added:

- How the annotation text strings will be defined based on attributes in the linked feature class

- Which features in the linked feature class will be annotated by the annotation class

For example, if you have an annotation feature class for cities, you could have annotation classes of varying text sizes and scale ranges for small, medium, and large cities—all managed within a single annotation feature class. Annotation classes save you from having to define and maintain multiple annotation feature classes.

Feature-linked messaging

Annotation links to features through a composite relationship with messaging. The feature class being annotated is the origin class in the relationship, and the annotation feature class is the destination class. As with other composite relationships, the origin feature controls the destination feature. If an attribute value for the origin feature changes, the linked annotation that is based on this attribute will automatically update to reflect the change. When the origin feature is moved or rotated, the linked annotation also moves or rotates with it. When an origin feature is deleted from the geodatabase, the linked annotation feature is also deleted.

An annotation feature class can be linked to only one feature class, but a feature class can have any number of linked annotation feature classes.

A feature-linked annotation feature class inside a feature dataset should link to a feature class within the same dataset. Similarly, stand-alone feature-linked annotation feature classes should link to stand-alone feature classes in the same geodatabase.

Comparing annotation options

Depending on your cartographic goals and the control you want over text placement, your options range from dynamic labeling, which is easy and quick, but with limited placement control, to standard and feature-linked annotation, which can be bulk-loaded from other feature classes and edited to the desired cartographic appearance and placement. The table below summarizes the benefits and costs of the three main options for annotating features.

Yet another annotation option is map document annotation. These are graphic text elements placed on a map. Map document annotation is not recommended for annotating features, but is used for adding notes or explanatory text on a map.

	Feature-linked annotation	Standard annotation	Dynamic labels
Definition	Annotation class joined in composite relationship with feature classes.	Stand-alone annotation class.	Text properties, such as font and size, and a labeling field, defined in an ArcMap layer.
Framework	Geodatabase data model.	Geodatabase data model.	ArcMap document property.
Typical uses	Fine cartographic placement of annotation with attribute-based text.	Fine cartographic placement of annotation with fixed text.	When text values are attribute values that can be placed automatically on the map.
Composite objects	Yes	No	No
Referential integrity	Yes	No	No
Pros	Manages referential integrity and messaging behavior. Edited via ArcMap attributes inspector.	No editing overhead; can cross workspace and data source type.	No editing overhead; can cross workspace and data source type; can be used for SQL queries, labeling, and symbology.
Cons	Incurs editing overhead; must be defined only between tables in same geodatabase; still requires joins for SQL query, labeling, and symbology.	No referential integrity; no messaging; no support for many-to-many cardinality; still requires joins for SQL query, labeling, and symbology.	No referential integrity; no messaging; no support for many-to-many relationships.
Recommendations	Use feature-linked annotation for text, such as street names, building numbers, and gauges.	Use simple annotation for static text or text describing large or indeterminate areas.	Use dynamic labeling for quickly placing descriptive text, such as addresses or measurements.

A geographic information system spatially integrates information. When you design relationships among features, use the GIS to first discover spatial relationships, such as touch, overlap, intersect, and cover. When your implementation cannot use spatial relationships, you can use several ways in ArcGIS to discover relationships among features, such as geometry snapping during editing, joins, relates, and relationship classes.

There are many ways to associate features with each other in a geodatabase. Your job as a data modeler is to find the method that represents the behavior associated with your data and uses the spatial integrative powers of the GIS.

Modeling with spatial relationships

Your first choice for modeling relationships is to use the GIS to manage the spatial relationships inherent among features. Here are three examples:

- You can model the components of an electrical system on a network, using edges for primary circuits with secondary taps and junctions for transformers, poles, and devices. To find an associated feature, do a trace on the network. A network is a set of relationships among edges and junctions.

- You can manage the nested geography in census mapping data with topology. Topology rules can be defined to enforce spatial relationships, such as how census blocks nest into block groups, and block groups nest into census tracts. Topology rules maintain the topological integrity of your data.

- Some features can be associated by proximity, containment, or other relationship. You can use geospatial operators to aggregate features within other features and summarize feature values.

Modeling with attribute relationships

There are also many associations that require attribute relationships to be defined. You can have an association between a geographic feature, such as a parcel of land, and a non-geographic entity, such as one or more parcel owners.

You may also need to capture relationships among features that may be close in proximity but for which there is ambiguity. For example, a transformer may serve electrical power to several buildings, but unless you have secondary lines mapped, you can't have a clean, unambiguous association between a transformer and the buildings it serves.

For these two general cases, you have three additional choices to make about representing that association. Relationship classes let you easily navigate related features and objects in ArcMap. On-the-fly relationships (relates) and joins are used to optimize editing and drawing performance.

spatial relationships

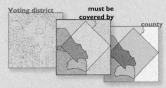

use a geodatabase topology to enforce spatial integrity among sets of features

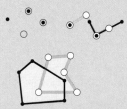

use spatial relational operators to test whether features touch, contain, or intersect

use parcel fabrics to manage spatially integrated sets of features such as land parcels

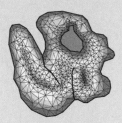

use terrain datasets to model surface relationships

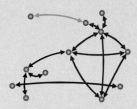

use networks to model systems of connected line features

attribute relationships

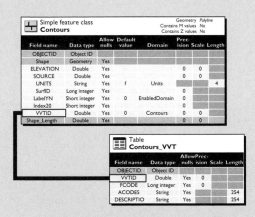

use joins and relates to navigate simple attribute relationships

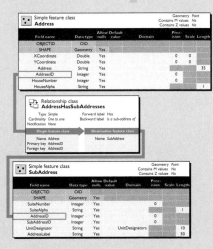

use relationship classes to persist complex and attributed relationships in the geodatabase

Geodatabases have data organized into multiple feature classes and tables. This is a good practice because it prevents information from being duplicated in several places. But you will often want to access information linking a feature class to another feature class or table. When you need to access information in this way, you can link tables to each other.

Features in feature classes and rows in tables can have attribute relationships established between them through a common field called a key. A key is an attribute of a table that uniquely identifies each record in that table. ArcGIS allows you to associate records in one table with records in another table through the values in their key fields. The key field in the origin class of a relationship is called the primary key while the key field in the destination class is called the foreign key. You can make these associations in several ways, including by joining or relating tables temporarily in your map or by creating relationship classes in your geodatabase that maintain more permanent associations.

The example below shows how a feature class and table are linked through foreign keys through a one-to-one relationship.

A feature class with voting districts is linked to election results in a table. The primary key in the District table is a field called Precinct. The foreign key in the Election table is a field also called Precinct. By linking these tables through matched key values, you can produce a map displaying and analyzing election results.

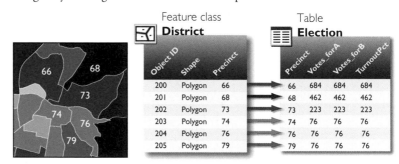

While the relationship between voting districts and election results is one-to-one, it would not be a good practice to combine these values into a single feature class. The reason is that voting districts have a life span over multiple elections. It's a better practice to create a new table for each election and link the feature class and tables when you want to study that election result. You can link many tables to a feature class and this gives you considerable modeling freedom.

The example below shows how keys are used to model a many-to-one relationship. The key field is a land-use code, which identifies a type of land use for each parcel of land.

A feature class with land parcel is linked to land use information in a table. The primary key in the Parcel feature class is a field called LandUse. The foreign key in the LandUse table is a field called LU_Code. This is a many-to-one relationship because there are many parcels which share a land-use code.

Tables with auxiliary information like LandUse are often called "lookup tables". You can use them for specifying additional information about categories of features without adding duplicate information in the main feature class.

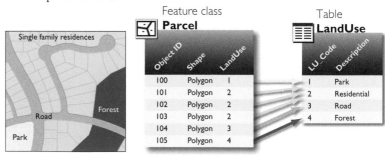

As you can see, linking tables through key fields reduces redundancy of data and is a central concept for managing databases. Once tables are linked—either by relates, joins, or relationship classes—you can work with them in ArcGIS as though these tables have been combined into one.

Relates

Relating datasets through a relate in ArcGIS defines a relationship between two datasets based on a key. The relate is established at the layer level and can be used in the mapping aspects of ArcGIS to navigate relationships without modifying the underlying data. Relates do not append the attributes of the origin dataset to the destination dataset; instead, the relate acts as a pointer between related datasets and can be used to access data when necessary.

Joins

Joining datasets in ArcGIS will append the attributes from one table onto the other based on a field common to both. When an attribute join is performed the data is dynamically joined together, meaning nothing is actually written to disk.

Relationship classes

A relationship class stores information about associations among features and records in a geodatabase and can help ensure your data's integrity. You may establish a relationship class between two tables, a table and a feature class, and two feature classes. Relationship classes provide many advanced capabilities not found in joins and relates. Relationship classes also have a series of properties that apply to all relationships in the relationship class. These properties define the behavior of the relationship class and are defined in greater detail in the following subsections.

Besides identifying the associated objects, relationship classes can have additional properties, such as type, cardinality, notification, and path labels.

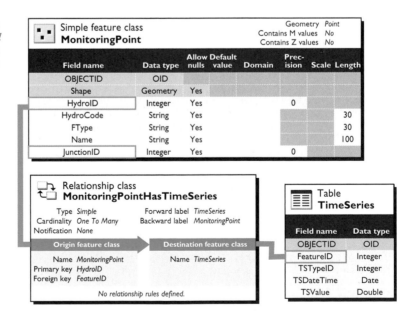

To link tables together by attribute values, you've seen that you have three choices: relates, joins, and relationship classes. Which method is right for linking two tables together depends on several criteria, such as the cardinality of the relationship (one-to-one, one-to-many, many-to-one, or many-to-many) and whether you want to control associations among features and rows when editing.

When to use relates

Use relates when you want to establish a one-to-many or a many-to-many relationship between a layer and a table. Unlike joins, relating tables simply defines a relationship between two tables and you can navigate through many destination rows in one-to-many and many-to-many relationships.

Relates are like simple relationship classes defined in the geodatabase, but relates offer improved editing performance.

When to use joins

Use joins when you want to establish one-to-one or many-to-one relationships between a layer's attribute table and the table containing the information you wish to join. Joins are not recommended for one-to-many or many-to-many relationships between a layer and a table because a join will only find a match in a table for the first joined feature in a layer.

Joins are best suited for labeling and symbology.

When to use relationship classes

You should use joins and relates whenever possible, but for some situations, relationship classes offer several important advantages.

Relationship classes help ensure referential integrity. For example, the deletion or modification of one feature could delete or alter a related feature. Furthermore, a relationship class is stored in the geodatabase, which makes it accessible to anyone who uses the geodatabase.

Relationship classes support all cardinalities—one-to-one, one-to-many, and many-to-many—and may have attributes about the relationship itself.

A relationship class can be set up so when you modify an object, related objects update automatically. This can involve physically moving related features, deleting related objects, or updating an attribute. For example, you could set up a relationship such that whenever you move a utility pole, attached transformers and other equipment move with it.

By setting relationship rules, a relationship class can restrict the type of relations that are valid. For example, a pole may support a maximum of three transformers. A steel pole may support class A transformers but not class B transformers.

Relationship classes help you access objects while you're editing. You can select an object and find all related objects. Once you have navigated to the related object, you can edit its attributes. Regardless of how deeply chained, all the related classes are available for editing.

Because relationship classes are stored in the geodatabase, they can be managed with versions. Versions allow multiple users to edit the features or records in a relationship at the same time.

	Relates	Joins	Relationship classes
Scope	Cross workspace or data source	Cross workspace or data source	Geodatabase
Framework	Defined in map layer	Relational database/SQL	Geodatabase data model
Typical uses	Editing with low overhead	Labeling, symbology	Modeling compound objects
User interface for editing	VBA application in ArcMap	SQL queries	ArcMap
User interface for navigating	ArcMap	SQL queries	ArcMap
Composite objects	No	No	Yes
Referential integrity	No	No	Yes
Messaging	No	No	Yes, used for class extensions
Attributes	No	No	Yes
Relationship rules	No	No	Yes, by subtype
Cardinality	One-to-many, many-to-many	One-to-one, many-to-one	Up to many-to-many
Pros	No editing overhead; can cross workspace and data source type.	No editing overhead; can cross workspace and data source type; can be used for SQL queries, labeling, and symbology.	Manages referential integrity and messaging behavior. Edited via ArcMap attributes inspector.
Cons	No referential integrity; no messaging; no support for many-to-many cardinality; still requires joins for SQL query, labeling, and symbology.	No referential integrity; no messaging; no support for many-to-many relationships.	Incurs editing overhead; must be defined only between tables in same geodatabase; still requires joins for SQL query, labeling, and symbology.

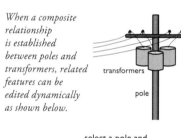

Relationship class **AddressHasSubAddresses**	
Type Simple	Forward label Has
Cardinality One to one	Backward label Is a sub-address of
Notification None	
Origin feature class	**Destination feature class**
Name Address	Name SubAddress
Primary key AddressID	
Foreign key AddressID	

Besides identifying the associated objects, relationships can have additional properties, such as type, cardinality, notification, and path labels.

A relationship is an association between two objects or features. Relationships are organized into relationship classes. You may establish a relationship class between two tables, a table and a feature class, and two feature classes. Each relationship in a relationship class has the same origin class and destination class.

Simple and composite relationship classes

A relationship class can either be simple or composite.

In a simple relationship class the origin objects can exist independently of the destination objects.

A composite relationship class is the opposite of a simple relationship class. In a composite relationship class the destination objects behavior and existence is dependant on the origin object. In this sense, the origin objects have control over the related destination objects. Take the example of a pole and transformer feature class participating in a composite relationship class. The pole feature class represents the origin class so if any pole features are deleted, any related transformers will be deleted also.

Cardinality

All relationship classes have a cardinality that specifies the number of relationships that can be formed between an origin object class and destination object class. Examples of cardinalities are one-to-one, one-to-many, and many-to-many. Using the previous scenario, the relationship class between the pole and transformer feature classes would be one-to-many because there can be many transformers related to a single pole feature.

When a composite relationship is established between poles and transformers, related features can be edited dynamically as shown below.

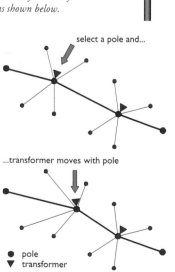

select a pole and...

...transformer moves with pole

- ● pole
- ▼ transformer
- ● meter
- ── primary line
- ── secondary line

Relationship rules

By setting rules, a relationship class can restrict the type of relationships that are valid. For example, an electrical pole may support a maximum of three transformers. A steel pole may support class A but not class B transformers. Relationship rules are a powerful way of maintaining data integrity by modeling complex real-world behavior between related objects.

Notifications

A notification is a message passed between the origin and destination classes when a significant event occurs, such as an edit or deletion. A relationship class may be used to propagate standard notifications between related objects. The notification direction property specifies these four notification options:

- No notifications are propagated.

- A notification is issued to the destination object only when the origin object is changed.

- A notification is issued to the origin object only when the destination object is changed.

- A notification is issued when either the origin or destination object is changed.

Attributed relationship classes

An attributed relationship class is a type of relationship class that uses a table to store the key values for each individual relationship. Because of this, attributes can be optionally added to each relationship. These attributes can be any qualities that describe the related objects. If the pole and transformer relationship class was attributed, values describing the mounting harness that is used to physically secure the transformer to the pole could be added to the table for each individual relationship.

Notifications are the mechanism that manages the lifespans of part objects based on the whole object in a composite relationship.

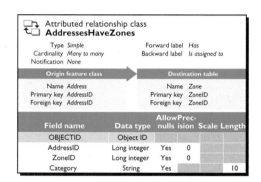

One-to-one relationships

One-to-many relationships

Many-to-one relationships

Many-to-many relationships

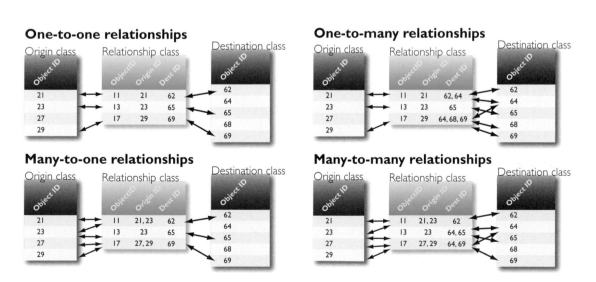

These are the cardinalities among origin and destination classes that you can model with relationship classes.

When you can't establish relationships among features through keys, you can use spatial joins. A spatial join lets you find associations among features like these:

- You can match features to the closest features.

- You can match features to features they are within.

- You can match features to features that they intersect.

When you perform spatial joins, ArcGIS uses a set of Boolean operators (called Clementini operators) that test the spatial relationships between two feature geometries. These operations can be applied to points, lines, and polygons.

The base geometry is the object invoking the operator. The comparison geometry is the object expressed as a parameter in the operator. The result of the relational operator is a Boolean value. No new geometries are created with this operator.

Contains

Does the base geometry contain the comparison geometry?

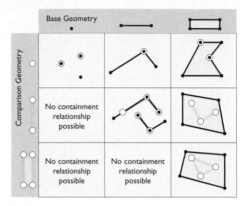

For the base geometry to contain the comparison geometry, it must be a superset of that geometry. A geometry cannot contain another geometry of higher dimension.

Equals

Does the base geometry equal the comparison geometry?

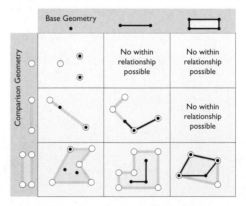

For the base geometry and comparison geometry to be equal, all of their constituent points must have identical coordinate values. The geometries that are compared must have the same dimension.

Within

Is the base geometry within the comparison geometry?

For the base geometry to be within the comparison geometry, it must be a subset of that geometry. A geometry cannot be within another geometry of lower dimension.

Crosses

Does the base geometry cross the comparison geometry?

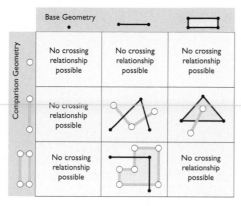

For the base geometry to cross a comparison geometry, they must intersect in a geometry of lesser dimension than the highest dimension.

Two lines can intersect at points. A line and an area can intersect at lines. There is no crossing relationship possible between a base area and comparison area. This is considered an overlap relationship.

Disjoint

Is the base geometry disjoint from the comparison geometry?

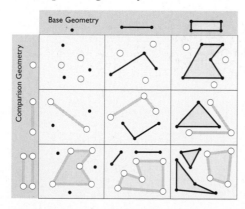

A base geometry is disjoint from a comparison geometry if they share no points.

Overlaps

Does the base geometry overlap the comparison geometry?

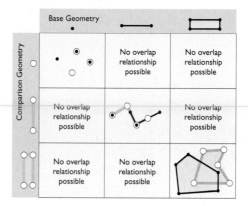

A base geometry overlaps a comparison geometry if their intersection is a geometry of the same dimension. An overlap relationship requires that both geometries be of the same dimension.

Touches

Does the base geometry touch the comparison geometry?

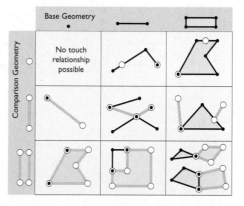

Two geometries touch when only their boundaries intersect.

Topology has historically been viewed as a spatial data structure used primarily to ensure that the associated data forms a consistent and clean topological fabric. With advances in GIS software development, an alternative view of topology has evolved.

The geodatabase supports an approach to modeling geography that integrates the behavior of different feature types and supports different types of key relationships.

In this context, topology is a collection of rules and relationships that, coupled with a set of editing tools and techniques, enables the geodatabase to more accurately model geometric relationships found in the world.

Features exist in a tapestry, with many types of spatial relationships. A topology models these spatial relationships, both among features within a single feature class and between features in two separate feature classes.

Uses of topology

Topologies determines how point, line, and polygon features share geometry. Topology is employed in order to:

- Constrain how features share geometry. For example, adjacent polygons such as parcels have shared edges and street centerlines, and census blocks share geometry.

- Define and enforce data integrity rules, such as no gaps should exist between polygons or there should be no overlapping features.

- Support topological relationship queries and navigate feature adjacency and connectivity.

- Support sophisticated editing tools (tools that enforce the topological constraints of the data model).

- Construct features from unstructured geometry such as constructing polygons from lines.

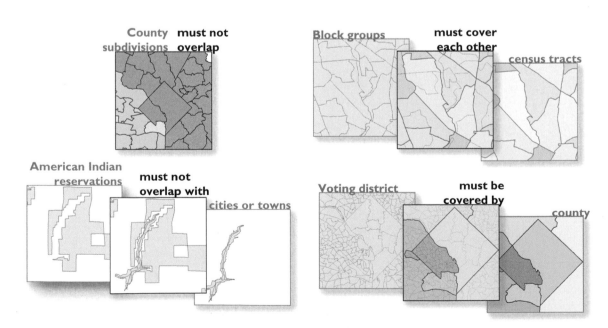

Sample topology rules applied to feature classes that comprise a data model for nested subdivisions of census administrative units.

How topology works

A topological data model represents spatial objects (point, line, and area features) as an underlying graph of topological primitives—nodes, faces, and edges. These primitives, together with their relationships to one another and to the features whose boundaries they represent, are defined by representing the feature geometries in a planar graph of topological elements.

Face	Edges	Nodes
A	1, 5, 6	1, 6, 5, 3
B	2, 4	1, 3, 2
C	8, 3, 4, 5	2, 3, 5, 4
D	7, 6, 8	6, 4, 5

Polygon features

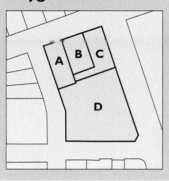

Topological elements

Edge	Left face	Right face	From node	To node
1	-	A	1	6
2	-	B	2	1
3	-	C	2	4
4	B	C	3	2
5	C	A	3	5
6	D	A	5	6
7	D	-	6	4
8	D	C	4	5

Properties of a geodatabase topology

The name of the topology to be created.

List of feature classes. First you need a list of the feature classes that will participate in a topology. All must be in the same coordinate system and organized into the same feature dataset.

The cluster tolerance used in topological processing operations. The cluster tolerance is often a term used to refer to two tolerances: the x,y tolerance and the z-tolerance. The default value for the cluster tolerance is 10 times the coordinate resolution.

Topology
ParcelFeatures_Topology   Cluster tolerance 0.000247

Participating feature classes and ranks

Feature class	Rank
Corner	1
Boundary	2
TaxParcel	3

Topology rules

Origin feature class	Topology rule	Comparision feature class
Corner	Must be covered by endpoint of	Boundary
Boundary	Endpoint must be covered by	Corner
Boundary	Must not have dangles	
TaxParcel	Boundary must be covered by	Boundary
TaxParcel	Must not overlap	

The relative accuracy rank of the coordinates in each feature class. If some feature classes are more accurate than others, you will want to assign a higher coordinate rank. This will be used in topological validation and integration. Coordinates of a lower accuracy will be moved to the locations of more accurate coordinates when they fall within the cluster tolerance of one another. Features with the highest accuracy should receive a value of 1, less accurate feature classes a value of 2, even less accurate feature classes a value of 3, and so on.

A topology has many topology rules. Some rules specify the valid spatial relationships among features in one feature class, other rules specify valid spatial relationships among features in two feature classes.

Once you have created a topology and loaded data, at any time you can run a validation against the feature class contents of the topology. Validate performs the following processing tasks:

- Cracking and clustering of feature vertices to find features that share geometry (have common coordinates)

- Inserting common coordinate vertices into features that share geometry

- Running a set of integrity checks to identify any violations of the rules that have been defined for the topology

- Creating an error log of potential topological errors in your feature dataset

Before validate	Clustering	Cracking	After validate

Cluster tolerance

● Inserted vertex
○ Existing vertex
○ Existing endpoint

It is important to note that the x,y tolerance is not intended to be used to generalize geometry shapes. Instead, it's intended to integrate linework and boundaries during topological operations, which means integrating coordinates that fall within very small distances of one another. Because coordinates can move in both x and in y by as much as the cluster tolerance, many potential problems can be resolved by processing datasets with commands that use the cluster tolerance. These include handling of extremely small overshoots or undershoots, automatic sliver removal of duplicate segments, and coordinate thinning along boundary lines.

Cluster processing

Creating topological relationships involves integrating the coordinate locations of feature vertices to make them coincident. This occurs among features in the same feature class as well as between the feature classes that participate in the topology.

A cluster tolerance is used to integrate vertices. All vertices that are within the cluster tolerance may move slightly in the validation process. The default cluster tolerance is the minimum possible cluster tolerance and is based on the precision defined for the dataset. The default cluster tolerance is 0.001 meters in real-world units. It is 10 times the distance of the x,y resolution (which defines the amount of numerical precision used to store coordinates).

When a vertex of one feature in the topology is within the xy tolerance of an edge of any other feature in the topology, the topology engine creates a new vertex on the edge to allow the features to be geometrically integrated in the clustering process.

When clustering feature vertices during topology validation, it is important to understand how the geometry of features is adjusted. All vertices of any feature class that participates in a topology can potentially be moved if they fall within the x,y tolerance of another vertex. Vertices of higher-ranking features will not move to lower-ranking features, but vertices of equal-ranked features will be geometrically averaged.

After your initial validation, which runs against all features in the feature dataset, subsequent validations can be run only against the changed areas in your feature dataset. Dirty areas are areas that have been edited, updated, or affected by the addition or deletion of features. Dirty areas allow the topology to limit the area that must be checked for topology errors during topology validation.

Original feature | Vertex added to feature | Dirty area is created

Two types of topology

You have two options for validating spatial relationships among features: geodatabase topologies, which give you a rich set of configurable topology rules, and map topologies, which make it easy to edit the shared edges of feature geometries.

A geodatabase topology lets you accurately model geometric relationships between features. By selecting a set of feature classes with a ranking and a set of topology rules between those feature classes, you can enforce geometric relationships, such as making sure that land parcels don't overlap and census blocks aggregate into block groups.

A map topology performs efficient topological editing through shared-edge editing tools. It has the advantage of letting you edit feature classes across many feature datasets, as well as shapefiles.

	Geodatabase topology	Map topology
Description	A geodatabase topology manages a set of feature classes that share geometry; is used to integrate feature geometry, validate features, control editing, and define relationships between features.	A map topology manages a set of simple feature classes that share geometry; is used to integrate feature geometry and control editing tools.
Scope	Feature classes in a feature dataset.	Feature classes in multiple feature datasets or shapefiles in a folder.
Definition	Object in a feature dataset with topology rules.	Created for duration of ArcMap session.
Rules	User sets any of several dozen topology rules.	Rules such as coincidence, covering, and crossing.
Validation	Rules evaluated when topology is validated.	Shared-edge validation applied during edits.
Reporting errors	Symbology for error shapes set in topology layer.	Errors cannot be created using the Topology Edit tool.
Correcting errors	User interface for locating and correcting errors.	Errors cannot be created using the Topology Edit tool.
Pros	Geodatabase topologies manage a set of rules and errors associated with the violations of those rules. They define valid spatial relationships between features.	Participant feature classes can be in different feature datasets. Map topologies can be used with shapefiles. They incur no editing overhead.
Cons	A geodatabase topology does not catch errors during feature creation, but during a validation procedure.	Map topologies have no rules or errors. They are defined during an ArcMap session, not in the geodatabase data model.
Recommendations	Use geodatabase topologies when you want to apply a set of your organization's topology rules.	Use map topologies when you want to perform quick shared-edge editing.

Features are displayed on a map with points, lines, and polygons.

Points represent small features on a map such as buildings and survey points. Lines represent narrow features such as streams and roads. Polygons represent broad features such as lakes and parcels of land.

How features are represented on a map depends on the map scale. A road in a transportation network will be drawn as a line. But a road in a subdivision map will be drawn as a polygon. How features are represented depends on the application and the range of map scales in which they are to be drawn.

Features in a geodatabase

Features are stored in a feature class. A feature class is a table with a special field for the feature shape.

Tables in a geodatabase work in the same way as tables in any relational database management system.

A table is organized into rows and columns.

A row is the fundamental unit of information in a table and comprises a collection of properties for an object.

A column represents common attributes for the row in a table, such as name, type, date, or other value. All rows have the same set of columns.

Fields specify the format of values within a column: number, text, shape, date, or other.

All tables in a geodatabase have a predefined field called ObjectID. ArcGIS manages this field to uniquely identify each row within each table in a geodatabase.

Tables can have a special field defined for subtypes. You can define rules for features and objects by subtypes. This helps you minimize the number of tables and feature classes for an application.

Feature classes and feature datasets

Feature classes can be stored either at the top level of a geodatabase or within a feature dataset. Feature datasets are used for the spatial integration of related feature classes, such as sets of feature classes that participate within a topology, network dataset, terrain dataset, or geometric network.

A property of a feature dataset is a spatial reference. This defines the coordinate system for all the feature classes in a feature dataset. If a feature class is not stored within a feature dataset, it has a spatial reference property. See Chapter 2, 'Coordinate systems and map projections'.

The shapes of features

There are five types of features: point, multipoint, line, polygon, and multipatch.

Feature locations are defined through one or a set of x, y or x, y, z coordinate values. Coordinate values for features are defined within a coordinate system.

Features have optional measure values (m-values) which are used for defining linear referencing on routes. See Chapter 5, 'Linear referencing with routes'.

Lines and polygons can be composed of one of four types of curves: line, circular arc, elliptical arc, or Bézier curve.

Line and polygon features can be single-part (have one curve) or multi-part (have multiple curves).

Describing attributes with fields

Fields describe the attributes in a table and can model diverse types of information such as identifiers, numeric data, descriptive text, feature shapes, and imagery.

Feature attributes can be validated through domains, which are lists of valid values.

Numeric fields representing integer values can be short integer (a range of about -32,000 to +32,000) or long integer (a range of about -2 billion to +2 billion). Numeric fields representing real values can be float (with about seven significant digits) or double (with about 16 significant digits).

Text fields are used for feature names and descriptions. They can be defined for any number of characters and they can also handle all the languages of the world with the Unicode system.

Feature classes have a special field called Shape which stores the geometry of each feature.

A raster field can be used for storing imagery such as photographs or scanned documents.

Validating features

You can apply behavior to features through default values, attribute domains, split-merge policies, connectivity rules, relationship rules, and topologies.

Default values assign a predefined attribute value when a new feature is created.

Attribute domains are used to ensure the validity of attribute values. There are two types of domains: range domains and coded-value domains. A range domain constrains numeric attribute values to be within a minimum and maximum value. A coded-value domain is a list of valid values and can be applied to text, numeric, and date values. Coded-value domains have a code value and a description.

Split-merge policies control what happens to specified attribute values when a feature is split or two features merged.

Connectivity rules constrain how network features can be connected to each other. See Chapter 4, 'Linear modeling with networks'.

Relationship rules allow you to restrict the type of objects in the origin feature class or table that can be related to a certain kind of object in the destination feature class or table

Topology rules enforce the spatial integrity of features, such as ensuring that features don't overlap.

Annotation

An annotation is a type of feature that provides a description of a place or feature. Annotation is stored in annotation feature classes.

Annotation can be standard and feature-linked. Standard annotation is not formally associated with features on the map and is used to label a place. Feature-linked annotation is associated with a specific feature and can display specified values from field values in a feature class.

Defining relationships among features

You can define or discover relationships among features and objects in a variety of ways.

If your features form a connected system, such as roads, streams, or utilities, you should manage these features within a geometric network or network dataset. See Chapter 4, 'Linear modeling with networks'.

If your features should follow spatial integrity rules, such as not overlapping, use a topology, which is a collection of topology rules governing valid spatial relationships among features in one or two feature classes.

Some relationships are defined through a key, a common field tying one feature or object to another. A key can be used to associate a parcel of land in a feature class to an landowner identified in a table.

There are three ways to establish relationships through key fields: relates, joins, and relationship classes.

A relate defines a relationship between two datasets based on a key field with common values for the related objects. A relate is defined within a map layer and does not modify the related tables.

When you work with a relate on a map, you can work with data from two tables or feature classes as if they were merged. Relates are like simple relationship classes and offer improved editing performance.

A join is similar to a relate but will append attribute values to the first table in a join. Joins are best suited for labeling and symbology.

A relationship class is a collection of relationships between two datasets in a geodatabase. Relationship classes allow greater control over related objects.

Testing spatial relationships

When you can't establish relationships among features using keys, you can use spatial joins.

With a spatial join, you can define relationships based on whether the geometries of two features are equal, containing, within, crossing, disjoint, overlapping, or touching.

Ensuring spatial integrity with topologies

A topology is a set of topology rules that are used to ensure the spatial integrity of features within a feature class or between two feature classes.

You work with a topology in an editing environment to find and correct features that don't meet your prescribed topology rules, such as ensuring no gaps or overlaps within a set of features.

You can also use a map topology to easily edit the shared edges of feature geometries within a map editing session.

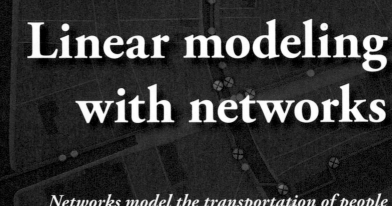

Linear modeling with networks

Networks model the transportation of people and resources such as water, electricity, gas, and communications. Networks constrain flow to edges such as streets and river reaches, which join at junctions such as intersections and confluences. The geodatabase has two core network models: the network dataset models transportation networks, and the geometric network models directed-flow systems such as river networks and utility lines.

4

Networks pervade our daily life. We drive cars from home to work on a street network, cook our dinners with natural gas delivered through gas utility lines, catch up with news and send emails through the internet, and visit relatives by flying on an airline route system.

Networks channel flow. Certain phenomena flow in a continuous field across a region, such as rainfall or temperature. But rainfall collects into streams, and nearly all resources that we process or goods that we manufacture flow in a constrained way, carried along networks of streets, pipes, cables, and channels.

A network is a one-dimensional system of edges that connect at junctions that transport resources, communications, and people. Three common types of networks are transportation, rivers, and utility networks.

Transportation networks

Transportation involves the movement of people and the shipment of goods from one location to another.

Streets are the ubiquitous network—everyone spends a fraction of every day traversing this network. Streets have two-way flow, except for situations such as one-way streets, divided highways, and transition ramps.

People make their daily travel optimal by hopping from one mode of transport to another, switching between walking, driving, riding a bus or train, and flying. People also use natural hierarchies in the transportation network. Trips of any distance usually begin by driving to the closest freeway on-ramp and proceeding to the off-ramp closest to the destination.

With GIS software, you can analyze a transportation network to support planning goals such as relieving congestion, mitigating pollution, optimizing delivery of goods, and forecasting demand for transportation.

Street, rail, and subway systems have well-defined geometry for the edges of the network, but transportation systems such as airline routes and shipping lanes have indeterminate or variable edges with geographic junctions at airports and harbors.

Some transportation tasks are:
- Calculating the quickest path between two locations
- Determining a trade area based on travel time
- Dispatching the closest ambulance to an accident
- Finding the best path and sequence to visit customers

The basic analysis done on river networks starts with estimating peak and average rainfall (from radar or model assumptions), determining how water gathers from catchment areas to river reaches, and how it accumulates downstream at confluences.

River networks

Rainfall on the landscape accumulates from rivulets to streams, rivers, and finally, an ocean. The shape of the surface directs water to a stream network. Gravity drives river flow from higher elevations to sea level.

A hydrologic network usually models a river as a connected set of stream reaches (edges) and their confluences (junctions). When a stream drains into a lake, hydrologic models continue the flow along an arbitrary line midway between shores until an outlet is reached.

Special large-scale hydrologic project models may include three-dimensional analysis of flow lines through a channel volume, but simplifying a river to a one-dimension line network is suitable for most applications.

Most parts of a river network form well-drained dendritic networks, with distinct channels and flow directions. In flat terrain, river flow becomes more complicated—a large river near an ocean often forms a delta with a complex braided network, and tidal effects can reverse flow near the shore.

Some hydrologic tasks are:

- Deriving catchments on a surface model for each stream reach
- Accumulating rainfall on catchments, transfer flow to reach
- Using gauge values, predict flood surge along a river
- Designing a system of channels and holding ponds for high water
- Managing diversion of water for agriculture or city water works

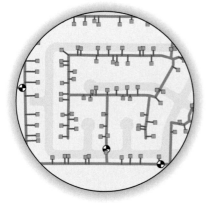

Utilities are concerned first with the safety of customers and employees, followed by the reliability of the system, and then cost efficiency in system operations. GIS is an effective tool to reach these goals.

Utility networks

Utility networks are the built environment that supplies energy, water, and communications and removes effluent and storm water. Water utilities are gravity driven or pressurized, depending on terrain. Flow in a gas utility is driven by pressure in pipes. Electric power flows from high voltage potential to low. Pulses of light carry communications in a fiber optic network.

Utility networks have a nominal flow condition, with a few sources delivering a resource to many points of consumption.

Some utility networks tolerate loops, such as a water network. For other utilities, a loop is a fault condition, such as an electrical short circuit. All utility networks contain dynamic devices such as valves and switches that can interrupt or redirect flow in the event of an outage or system maintenance.

Some utilities such as telecommunications and electrical networks have multiple circuits on a common carrier (edge), such as electric lines with three phases of power or twisted-pair lines in telephony.

Some utility network tasks are:

- Establishing the direction of a commodity flow
- Finding what is upstream of a point
- Closing switches or valves to redirect flow
- Identifying isolated parts of the network
- Finding facilities that serve a set of customers

Drawing networks involves a variety of cartographic techniques. Depending on your application, you will apply some of these and other techniques:

- Use point symbols for discrete objects on a network that are a source of flow, measure flow, interrupt flow, redirect flow, or consume flow.

- Use line symbols for the lines that channel flow, such as street centerlines, streams, and power lines. Apply a type classification to differentiate features on a map by hierarchal rank, such as freeways to dirt roads.

- Use fill symbols when drawing network lines at large scale, such as drawing road areas (between the curb lines) or a polygon representing the areal shape of a significant river.

- Draw how line features such as roadways in a highway interchange cross in correct stacking order.

- Label networks with names for linear features such as streets, metro lines, and streams.

- Use line end symbols such as arrowheads to display the flow status of your network.

- Highlight sections of a network where a key attribute such as traffic volume or electrical voltage reaches a critical stage.

- Draw optimal routes through a network, such as shortest drive time across city streets.

Like other geographic features, network features can change dimension as you zoom into a map. For example, a regional river network will be drawn as a set of lines at a scale of 1:1,000,000, but a map at scales such as 1:24,000 should begin to show the larger rivers with their areal shapes between shorelines.

Techniques for roads

Like other geographic features, network features are drawn with point, line, and fill symbols by a type classification. These are sample symbols and types for hydrography layers on maps.

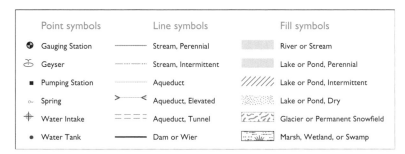

A large-scale street map has several special cartographic requirements. By custom, streets are drawn with casings, which symbolically represent the curb line (or edge) of a road. Also, the natural hierarchy of roads, spanning dirt roads, city roads, major roads, and divided highways, needs to be illustrated using colors, patterns, and line thicknesses. Road names and highway shields are labeled. Finally, where roads cross each other or objects, such as rail lines and rivers, the correct stacking order of the objects needs to be drawn.

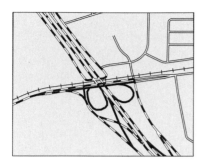

This road map shows the use of many finely defined layers in groups using elevation levels to show the correct order of roadways.

Drawing clean road casings is easy. Define two layers on the same street feature class. On the bottom layer, set the line symbol to be the casing color, usually black, with a line thickness of three points, or so. On the layer above, draw the fill with a lighter color such as white and a narrower thickness, about two points.

To draw a sophisticated road map like that on the left, you need a street feature class with two attributes: a type classification and an elevation level. To achieve fine control of stacking order, you need to precisely sort the drawing order of features. One way to do this is to create many layers on the street feature class and draw them in a controlled order.

The idea is to first draw feature casings (thicker, darker lines) by type for ground level. Then draw street fills (thinner, lighter lines) by type for ground level. And repeat for each level. To organize these many layers, use group layers to collect layers by elevation. For more information on high-quality rendering of transportation and other layers, see Chapter 8, 'Cartography and the Base Map', in *Designing Geodatabases: Case Studies in GIS Data Modeling* by ESRI Press.

Schematics

Depicting the connectivity of a network is often the most important part of a map. In this view, geographic detail is unimportant.

Situations calling for schematics range from passengers who want a simple, clean map for navigating a subway system to power engineers who need to make control decisions regarding the operation of an electric system.

A schematic is a simplified representation of a network that emphasizes connectivity over geographic shape. A schematic may preserve some geographic representation, such as keeping elements in relative geographic orientation, but detailed shape information is not shown. Some schematic graphic elements have no relation to geographic shapes; they show relationships along a network.

You work with schematic diagrams on the map in the same way as you work with charts or reports. A schematic is another graphical representation of a query against a database; it is composed of a set of schematic elements which have a graphic x,y location, maintain a link back to their original feature, and store symbology and type information.

This link back from a schematic element to the original feature is essential to automating schematics. In past practice, schematics were generated as one-time graphical products manually regenerated over time. Now, you can automate and refresh schematics in a geodatabase when the source network features are edited.

Generating schematics requires an investment in training and configuration because the available types and properties of schematics is rich and complex. That investment can pay off for organizations such as utilities by reducing manual editing and improving clarity and accuracy of schematics.

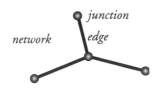

A network is comprised of edges which connect at junctions.

Networks are simple. They are comprised of two fundamental components, edges and junctions. Examples of edges are streets, transmission lines, pipes, and stream reaches. Examples of junctions are street intersections, fuses, switches, service taps, and the confluence of stream reaches.

Edges connect at junctions, and the flow from one edge—automobiles, electrons, and water—is transferred to another edge.

To analyze networks, geographers use a branch of mathematics called graph theory. Graph theory keenly interests us because it provides rigorous theorems concerning properties of networks and algorithms to solve network traversing problems. For example, one set of related algorithms for solving a problem in graph theory known as the Chinese Postman Problem provides optimized solutions for many applications such as postal delivery, street sweeping, bus passenger pickup and drop-off, and garbage collection.

The first graph

Graph theory is the foundation to our understanding of networks and topology. A GIS modeler should be familiar with the concepts and terminology of graph theory because it helps us to classify and model connectivity and adjacency relationships among geographic features.

The modern study of networks began in 1736 with an important mathematical result published by the Swiss mathematician Leonhard Euler. He considered a problem that was a local curiosity among the citizens of Königsberg in former East Prussia. The riddle was this—given seven bridges that cross a river and two islands, was it possible to take a walking tour across each bridge exactly once?

Euler solved the problem by thinking about the bridges and land masses as parts of a graph, an abstract set of junction objects (for land masses) and edge objects (for bridges). From this, he devised an easily understandable theorem that starts with an observation that the junction (or land mass) where you start a tour must be connected to an odd number of edges (bridges). This must be so because otherwise you must cross the same bridge twice to walk further. Likewise, the junction where a tour ends must also have an odd number of edges. Because you cannot have more than one start and one end to a tour, his theorem proved that any graph with more than two junctions connected to an odd number of edges cannot be traversed by traveling through each edge exactly once.

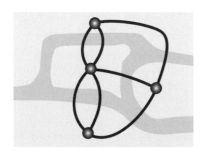

Euler modeled the seven bridges of Königsberg with four junctions for the land masses and seven edges for the bridges. This began the mathematical study of networks. GIS software uses results from graph theory to implement solutions to many practical routing problems.

Why is this piece of mathematical history interesting? Not because it's a clever solution to a curious problem, but because Euler's insight illustrates that when you abstract a geographic system to its underlying graph, you have a useful model for understanding the connectivity of that system.

Today, we do the same sort of abstraction when modeling networks. Or rather, the core network models in GIS software do much of this abstraction for us, leaving modelers with a set of network properties for fine-tuning exactly how a graph is derived from geographic features. This underlying graph enables us to perform network analysis such as finding the shortest route. Network trace solvers in GIS software implement many algorithms from graph theory.

Whenever you do something interesting with a network in GIS software, you are interacting with its underlying graph. GIS software manages bidirectional links between network features and elements in a graph so that you can naturally interact with network features to solve useful problems.

Graphs and their elements

Unfortunately, terminology is not consistently applied in the literature for graph theory. You will find the terms vertex or node variously used for intersections in a graph and arcs, chains, or links for the paths in a graph. We use the corresponding terms junction and edge for both the geographic network features and its graph elements.

Graph theory applies to a range of phenomena, far beyond geographic applications. This is a brief, non-rigorous description of graph concepts. Understanding graphs and their properties will help you make network modeling decisions.

A surface can be considered an infinite set of points on which you can sample continuous varying phenomena, such as elevation, precipitation, temperature, soil moisture, and spectral reflectance at many wavelengths.

Other phenomena are not continuous across a surface. People, manufactured goods, water, energy, communications, and commodities are transported in a constrained way, channelized by streets, railroads, pipes, fiber optic lines, and shipping routes. The set of paths and where they join is a network, a subset of the infinite points on a surface.

Graphs can represent systems that move along natural or man-made linear networks. For example, the interstate highway system can be abstracted to a set of edges representing sections of the interstate highway between interchanges and junctions representing the interchanges at cities.

A graph is the set of edges that connect at junctions. Two junctions span an edge. A junction can connect to one or many edges.

Properties of a graph

A graph can be directed or undirected. (A directed graph is also called a digraph.) Directed graphs model systems where each edge has a fixed direction of flow, such as a river network flowing downstream inside hydrologic channels. Undirected graphs model systems with no preferred direction of flow, such as street networks which predominantly allow travel in both directions.

Three important qualities of a graph pertinent to geographic networks are whether the network has directed flow, whether edges can cross on a plane or not, and whether junctions and edges form cycles inside a graph.

A graph can be planar or nonplanar. Planar graphs connect on a two-dimensional plane at junctions, without any crossing edges. Nonplanar graphs have edges that cross. On a street network, nonplanar graphs can model a bridge over a road. Geographic data commonly comes in planar and nonplanar form. Planar data allows you to use elevation-level information to model crossing streets. This is discussed in a later topic on elevation fields.

A graph can be cyclic or acyclic. A cycle (or circuit) is a set of connected edges that eventually return to a junction (in the case of directed graphs, a further condition is that the edges line up in flow order to close a cycle). A graph without cycles is called a tree graph (or acyclic graph) in graph theory. Examples are local area communication networks and river systems. A graph with cycles is called a cyclic graph. Application examples are streets and water utilities.

Knowing whether a graph is directed or undirected is the most important graph property when considering GIS network models. These two types of graphs are optimally stored with different internal data structures and have separate classes of algorithms for network analysis. Consequently, the geodatabase model has two fundamental representations for a network, which we discuss next.

The geodatabase implements two core models for representing networks: geometric networks and network datasets. Geometric networks are designed for directed networks such as an electric utility or a river network. Network datasets are optimized for undirected networks, such as road networks.

Network datasets and geometric networks contain geographic features in feature classes from which a graph is derived. The elements of this graph are stored in a logical network, a set of element and connectivity tables. There are differences in how features are defined in the two models and how operational workflows refresh the graph, but both involve the same conceptual abstraction of a graph from geographic features.

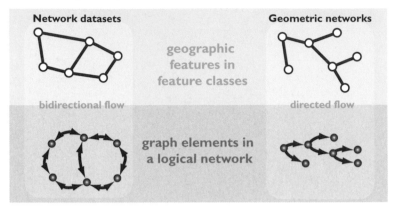

Geometric networks

Geometric networks are composed of two main elements: edges and junctions. An edge is a feature that has a length through which some commodity flows. A junction is a feature that allows two or more edges to connect and facilitates the transfer of flow and resources between edges.

Edges are further divided into two types: simple and complex. Simple edges allow resources to enter one end of the edge and exit the other end of the edge, such as a water lateral in a water network. Complex edges allow resources to flow from one end to the other, just like simple edges, but they also allow resources to be siphoned off along the edge without having to physically split the edge feature, such as a water main in a water network.

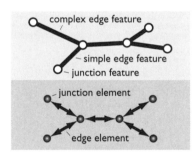

In this simple geometric network, one complex edge and two simple edges are represented with five edge elements.

Edges and junctions in a network are topologically connected to each other—edges must connect to other edges at junctions, and the flow from edges in the network is transferred to other edges through junctions. With geometric networks, connectivity is updated whenever you add or remove any network feature. When you edit features in a geometric network, the logical network is continuously updated.

Geometric networks have connectivity rules that allow you to specify which network features can properly connect to each other. Analytical solvers use weights (impedances on edges) to perform tracing functions useful to utility and hydrographic applications. Solvers include downstream trace, upstream trace, isolation trace, and path trace.

Network datasets

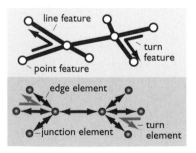

Network datasets are made from simple line, point, and turn features. Connectivity properties fine-tune how elements are discovered from features.

Network datasets are more analogous to a geodatabase topology—a network dataset has network sources defined on simple features. These features undergo no modification and can also participate in a topology. With network datasets, you can update connectivity on demand with a rebuild of the entire network.

Network datasets better model undirected networks. They allow flow in any direction and employ turns to model restrictions.

Network datasets have an attribute environment that uses costs, descriptors, restrictions, and hierarchy. Network datasets are optimized for large transportation networks.

This table summarizes and compares the two network models.

	Network dataset	Geometric network
Application	For transportation modeling	For utilities and natural resources modeling
Analysis	Pathfinding and allocation operations	Network tracing functionality
Sources	A network dataset is built from junction, edge, and turn sources, which are simple feature classes in a geodatabase feature dataset or shapefiles.	A geometric network is initially built from simple line features, from which simple edge, complex edge, and junction feature classes are made.
Features	Network sources are simple point and line feature classes. Some feature classes build the junction-edge connectivity model and other feature classes define turns, used to restrict the traversability of the network.	Features in a geometric network are in one of these network classes: junction, simple edge, and complex edge.
Connectivity	Connectivity in a network dataset is non-reactive and refreshed with a network build.	Connectivity in a geometric network is maintained on-the-fly after each edit; it does not have to be manually built.
Turns	Turns are line features that follow two or more connected street lines and model traffic scenarios such as restricted left or U-turns.	Geometric networks do not model turns.
Topology	Network source feature classes in a geodatabase can participate in topologies.	Geometric network feature classes cannot participate in topologies.
Attributes	Network datasets have an attribute model with costs, descriptors, restrictions, and hierarchy.	Geometric networks have a simpler attribute model using weights.
Workflow	Network connectivity is built on demand, similar to validating topology. Once any source feature class is edited, the network is invalid and must be rebuilt.	Network connectivity is continually maintained with every edit.
Multimodal modeling	A network dataset can model multimodal systems using connectivity groups.	A geometric network uses complex edges for a simple hierarchy and is not suited for multimodal systems.

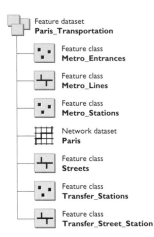

Feature dataset
Paris_Transportation

Feature class
Metro_Entrances

Feature class
Metro_Lines

Feature class
Metro_Stations

Network dataset
Paris

Feature class
Streets

Feature class
Transfer_Stations

Feature class
Transfer_Street_Station

A network dataset is defined as a set of feature classes in a common feature dataset, shapefiles, or StreetMap data.

The network dataset is the core geodatabase network model for representing undirected networks, particularly transportation networks.

Network datasets encapsulate decades of GIS modeling experience on optimizing street datasets for network analysis. You can see this in the network dataset's attribute model representing cost impedances, connectivity rules and groups, hierarchy, elevation fields, one-way restrictions, and turns. This model enables sophisticated analysis such as optimal routing and allocation, even on multimodal systems such as combined street, subway, and pedestrian traffic.

Sources for a network dataset

A network dataset contains network elements built from features in network sources. These sources are either feature classes in a geodatabase feature dataset, shapefiles, or feature classes in StreetMap format.

A network dataset in a geodatabase workspace can have zero or more junction sources (point feature classes), zero or more edge sources (line feature classes), and zero or more turn sources (line feature classes with turn attributes). The network dataset must have at least one junction source or one edge source.

A network dataset in a shapefile workspace has one edge source (line shapefile) and zero or more turn sources (line shapefiles with turn attributes). A network dataset in a StreetMap workspace has one or more edge sources (StreetMap line feature classes) and turns are modeled with topological associations among junctions and edges.

Making network elements

Network elements are made from features when a network dataset is built. You discover network elements and their connectivity by finding geometric coincidences of points, polyline endpoints, and polyline vertices. The network elements and connectivity information are stored in the logical network, a set of element and index tables inside a network dataset.

network features   network elements

polyline   edge

endpoint   junction

turn   turn

point

When built in a network dataset, a set of point and polyline features from many feature classes becomes a set of junctions, edges, and turns in the logical network.

Edges are network elements that connect to junctions. Edges are the links over which resources flow. Each edge has exactly two junctions. Junctions connect edges and facilitate navigation. A junction may be connected to one or more edges. When a network dataset is built, a system junction feature class is created with features at all unoccupied junctions. Turns record information about a sequence of two or more connected edges. Turns model turn restrictions such as no left turns or turn impedances.

This is how the three types of network elements are derived.

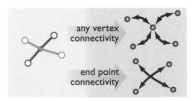

any vertex
connectivity

end point
connectivity

The edge connectivity policy can create two distinct graphs from two polylines that intersect at a vertex.

- Junction elements are created by points from junction sources that are coincident with polyline endpoints. Junction elements are created for all the remaining polyline endpoints also.
- Edge elements are created by polylines from edge sources. Most often, one edge element is made from one polyline feature. But depending on endpoint and vertex coincidence and whether endpoint or any vertex connectivity is set for the edge source, a polyline feature may be split into more than one edge element, with additional junction elements as well.
- Turn elements are created from turn features. A turn feature is a line feature with turn attributes that overlays two or more connected polylines.

A planar graph does not have any crossing edges without a junction at their intersection.

A nonplanar graph allows edges to cross without junctions, useful for modeling overpasses and tunnels.

Network elements do not have any geometry but combine to form a nonplanar network graph. Streets generally meet at surface level, so the transportation system is mostly planar. When roadways split elevations, such as at bridges or tunnels, this can be modeled by using nonplanar features for the overlapping lines or planar features with elevation fields marking the elevation levels at both ends of a line. While network datasets accommodate nonplanar networks, you will find that much commercial street data is mostly or completely planar.

Network connectivity

You can use connectivity groups to model multimodal systems, such as combined street and subway systems. Elements in different connectivity groups cannot connect except at shared junctions such as subway stations.

For junctions, a connectivity policy sets whether a point for a junction must be located at an endpoint or whether it can be at any mid-span vertex along a line. For edges, a connectivity policy determines whether junctions are added when vertices from different lines coincide or whether junctions are added only at line endpoints.

Attaching network attributes

Network elements have network attributes that are assigned, calculated, or derived from source feature attributes. Default values can be assigned to network attributes. Network attributes are used by network analysis solvers to enhance connectivity, define impedances, and improve solver performance. Network cost attributes can be accumulated during an analysis solution, such as accumulated travel time costs on edges visited along a route.

When creating a network dataset, you can identify special attributes from network sources with information about hierarchy (road classification such as highway, major road, minor road levels), elevations (setting the physical levels of roadways, like stories in a building), one-way or vehicular restrictions (such as no trucks or no pedestrians), and cost of travel time or distance traveled. With variations on implementation, commercial data vendors include these and other attributes on street networks.

Working with network datasets

A network dataset operates like a geodatabase topology—a connectivity graph is built on demand from point, endpoint, and vertex coincidence of participating feature classes.

A network source can participate in one network dataset. Any operations on a network dataset will not affect the network source feature classes. As simple feature classes, network sources from a geodatabase workspace are available to participate in topologies.

All sources in a network dataset must share a common spatial reference to ensure the spatial coincidence of vertices and endpoints. This is enforced when you add sources.

Geometric network feature classes (junction feature, simple edge feature, and complex edge feature) cannot participate in a network dataset.

Junction element table

Feature class	Feature ID	Element ID	User-defined network attributes		

Network attributes are added as columns to the element tables in the logical network. The junction, edge, and turn element tables have the same set of network attributes. The edge element table has network attributes for the forward and backward direction of the edge.

This simple street network shows how features from network sources combine in a network dataset to create a set of network elements whose connectivity is discovered and stored in the logical network, an internal set of network index tables.

This network dataset has one line feature class as an edge source, one turn feature class (a line feature class with turn attributes and features representing turns) as a turn source, and no point feature classes for a junction source.

Building a network dataset creates and adds a point feature class with computed junctions to the network. These are called system junctions.

Network datasets can have any number of edge sources, junction sources, and turn sources, but this simple example lets you see how network elements are created and how the important connectivity relationships are kept in the logical network.

In this example, each street feature matches one edge element, and this is typical in street data. Three conditions lead to the creation of multiple edge elements for one line feature: multipart polylines, any vertex connectivity between an edge and another edge, and mid-span connectivity between junction and a edge.

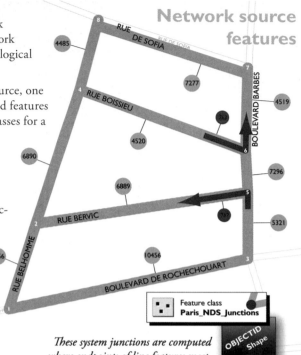

Network source features

These system junctions are computed where endpoints of line features meet. When a network dataset is built, system junctions are created where no other point sources are input.

Feature class **Paris_NDS_Junctions**

OBJECTID	Shape
1	Point
2	Point
3	Point
4	Point
5	Point
6	Point
7	Point
8	Point

The ten road lines in this simple street feature class join at their endpoints and contain attributes such as one-way restriction and hierarchy. In this example, the street network is planar. This means that no lines cross without a junction at the intersection.

Feature class **Streets**

OBJECTID	Shape	FULL NAME	METERS	FT MINUTES	TF MINUTES	DISP CODE	FUNC CLASS	Oneway	Hierarchy
4485	Polyline	RUE BELHOMME	25.38	0.076	0.076	40	5	T	5
4519	Polyline	BOULEVARD BARBES	29.01	0.087	0.087	30	2		2
4520	Polyline	RUE BOISSIEU	62.42	0.187	0.187	40	5	T	5
5321	Polyline	BOULEVARD BARBES	23.25	0.069	0.069	30	2		2
5356	Polyline	RUE BELHOMME	31.49	0.094	0.094	40	5	T	5
6889	Polyline	RUE BERVIC	75.62	0.226	0.226	40	5	T	5
6890	Polyline	RUE BELHOMME	49.21	0.147	0.147	40	5	T	5
7277	Polyline	RUE DE SOFIA	55.41	0.166	0.166	40	5	T	5
7296	Polyline	BOULEVARD BARBES	14.52	0.043	0.043	30	2		2
10456	Polyline	BOULEVARD DE ROCHECHOUART	86.22	0.258	0.258	30	2	T	2

These two turns have two edges and are used to model two prohibited left turns.

Feature class **Turns**

OBJECTID	Shape	Edge1End	Edge1FCID	Edge1FID	Edge1Pos	Edge2FCID	Edge2FID	Edge2Pos	Edge3FCID	Edge3FID	Edge3Pos	AltID1	AltID2	AltID3
763	Polyline	N	13	4520	0.25	13	4519	0.25				56237813	56237812	
767	Polyline	Y	13	5321	0.75	13	6889	0.75				56237880	56237862	

Logical network elements

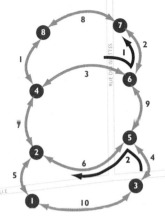

The logical network

When you build a network dataset, ArcGIS software calculates a logical network, which is a collection of connected junctions, edges, and turns but without any geometry or coordinates. Instead, a logical network is a set of tables that link feature classes and features to network elements and their connectivity. A logical network is a pure network graph stored in tables. These tables are regenerated and connectivity rediscovered each time you build a network dataset.

The internal details of the logical network are mostly hidden from the ArcGIS user. For example, you will never edit the junctions and edges directly, nor would you work with their IDs. Instead, you will continue to interact with and edit the features in the network sources. The junction and edge elements are regenerated quickly with a network build.

The purpose of the logical network is to provide the data structure necessary to enable network analysis such as route finding and service-area location.

Junction element table

Feature class	Feature ID	Element ID	User-defined network attributes
NDS_Junctions	1	1	
NDS_Junctions	2	2	
NDS_Junctions	3	3	
NDS_Junctions	4	4	
NDS_Junctions	5	5	
NDS_Junctions	6	6	
NDS_Junctions	7	7	
NDS_Junctions	8	8	

The three element tables link junctions, edges, and turns to their respective network source features. These tables also contain network attributes.

The junction connectivity table establishes how junctions and edges are connected by maintaining a list of adjacent edges and junctions for each junction.

Junction connectivity table

JunctionID	Adjacent edges	Adjacent junctions
1	5, 10	2, 3
2	5, 6, 7	1, 5, 4
3	4, 10	5, 1
4	1, 3, 7	8, 6, 2
5	4, 6, 9	3, 2, 6
6	2, 3, 9	7, 4, 5
7	2, 8	6, 8
8	1, 8	4, 7

Edge element table

Feature class	Feature ID	Element ID	User-defined network attributes	
			From-to direction	To-from direction
Streets	4485	1		
Streets	4519	2		
Streets	4520	3		
Streets	5321	4		
Streets	5356	5		
Streets	6889	6		
Streets	6890	7		
Streets	7277	8		
Streets	7296	9		
Streets	10456	10		

Edge elements contain bidirectional edge attributes. This means that any network attribute has two fields: one for the forward direction of the edge, the other for the backward direction.

The edge-turn reference table identifies which turns exist on an edge. There are similar index tables for retrieving turns and edges for a given junction.

Edge-turn reference table

Edge ID	IDs of turns on edge
1	<null>
2	1
3	1
4	2
5	<null>
6	2
7	<null>
8	<null>
9	<null>
10	<null>

Turn element table

Feature class	Feature ID	Element ID	User-defined network attributes
Turns	763	1	
Turns	767	2	

Turns typically have restriction attributes that restrict left, right, and U-turns, and cost impedances representing the additional cost of travel.

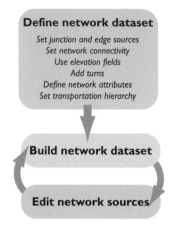

Define network dataset

Set junction and edge sources
Set network connectivity
Use elevation fields
Add turns
Define network attributes
Set transportation hierarchy

Build network dataset

Edit network sources

Once a network dataset is defined, a build process will generate the network elements which enable network analysis. Whenever a network source is edited, the network dataset must be rebuilt.

A network dataset can use a feature classification to make a three-level hierarchy, such as interstate highway, major road, and minor road. The logical network uses hierarchy to optimize route solving on large transportation networks.

Making a network dataset is like making a terrain dataset. A set of participating feature classes is defined, properties are set, and a graph is made. The properties of a network dataset affect how network elements are discovered from feature coincidence, refine the connectivity model, and optimize solver performance. These are the steps to creating a network dataset.

1   *Setting sources.* First, determine whether your network sources will be from a geodatabase feature dataset or a shapefile. If it is a geodatabase, you can add many point feature classes, line feature classes, and line feature classes with turn attributes as network sources.

2   *Adding turns.* Turns are added to the network dataset for one of two reasons: to restrict travel, such as no left turn allowed, or to add an additional travel time for the solver to perform that maneuver.

3   *Setting connectivity groups and policies.* You can set whether network sources should be organized into one or several connectivity groups. Connectivity groups are used to distinguish different transportation networks, such as subways and streets, that do not connect except at special junctions such as subway stations. You can also control how edges are built from lines. You may choose to create junctions wherever two lines have coincident vertices or create junctions at endpoints only. This choice involves whether your source features are planar or not and how you want to model crossing objects.

4   *Specifying elevation models.* You can use either z-coordinates on features or elevation fields to model crossing objects such as bridges, tunnels, and overpasses. An elevation field stores the logical levels of roadways, analogous to the stories in a building. Commercial data vendors commonly provide elevation fields on street networks.

5   *Specifying historical traffic data.* You can add traffic data to derive real time values for traversing lines.

6   *Defining network attributes.* You can add network attributes to network elements. All junctions, edges, and turns share the same attributes. Applications of network attributes include one-way restrictions, travel time, accumulated travel distance, and road classification.

7   *Specifying driving directions settings.* If you plan to generate driving directions text of your network analyses, be sure that your edge sources have fields containing the information needed to generate driving directions such as road class, street names, highway shields, and boundary information.

8   *Building the network dataset.* You will build the network dataset once the network dataset sources and properties are set. This will create the network elements in the logical network. The network must be rebuilt when a network source is edited. This works well with many organizations' data workflows because it is quick and easy to rebuild a network dataset.

TYPES OF SOURCE WORKSPACES

Network sources can reside in either a geodatabase feature dataset or shapefile workspace.

- Use sources from a geodatabase feature dataset for the greatest modeling flexibility and integration with other geodatabase data models such as topology.

StreetMap feature classes use a highly compressed form of spatial and attribute data. This format is used for data on the ESRI Data & Maps and StreetMap USA DVD.

- Use sources from a shapefile workspace for fast and convenient creation of network datasets. Because only one edge source is allowed, multi-modal networks are not possible.

This table compares the characteristics of the three types of network source workspaces.

	Geodatabase	Shapefile	StreetMap
Application	Network datasets in geodatabases give you a modeling framework with topologies, relationships, versioning, and rules.	Making network datasets from shapefiles is quick and easy. Also, much commercial street data is delivered in shapefile format.	The StreetMap format compresses vector data so that countrywide street networks can fit on a CD.
Data scope	All feature classes serving as network sources must be in the same feature dataset (and thus share a spatial reference).	A line shapefile and an optional turn shapefile must be in the same shapefile workspace (folder) and must have the same spatial reference.	StreetMap network datasets are built by data vendors with StreetMap feature classes. They cannot be modified by users.
Data maintenance	Feature classes can be directly edited.	Shapefiles can be directly edited.	StreetMap feature classes cannot be edited.
Junction sources	Multiple point feature classes from a feature dataset can be added as junction sources to a network dataset. Also, system junctions are calculated for all edge intersections not occupied.	System junctions are calculated at the intersections of edges. No junctions are input from shapefiles.	System junctions are calculated at the intersections of edges. No junctions are input from StreetMap feature classes.
Edge sources	Multiple line feature classes from a feature dataset can be added as edge sources.	Only one line shapefile can be an edge source. Junctions are built for this line shapefile.	Only one StreetMap line feature class can be an edge source. Junctions are built for this line feature class.
Turn sources	Multiple line feature classes with turn attributes from the feature dataset can be added to a network dataset.	Multiple line shapefiles with turn attributes can be added to a network dataset.	A simple turn restrictions table can be added to a StreetMap network dataset.
Topology	Feature classes that are network sources can also participate in a geodatabase topology.	Shapefiles are edited with tools applying simple topological relationships, but do not participate in a geodatabase topology.	A StreetMap network supports simple end-to-end topological relationships.

When you create your network dataset, you will make choices that determine which edge and junction elements are created from source features. Ensuring that edges and junctions are formed correctly is important for accurate network analysis results.

Connectivity in a network dataset is based on geometric coincidences of line endpoints, line vertices, and points and applying connectivity rules that you set as properties of the network dataset.

Connectivity groups

Connectivity rules begin with the definition of connectivity groups. Each edge source is assigned to exactly one connectivity group and each junction source can be assigned to one or more connectivity groups. Junctions that are assigned to two or more connectivity groups are the only way that edges in different connectivity groups can connect.

Connectivity groups are used to model multimodal transportation systems. For each connectivity group, select the network sources that interconnect. In the subway and street multimodal network example, metro lines and metro entrances are assigned to one connectivity group and streets and metro entrances to another.

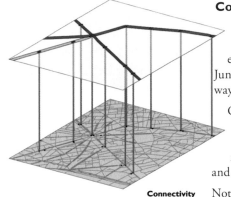

	Connectivity groups	
	1	2
Metro_Line	☐	☑
Street	☑	☐
Metro_Entrance	☑	☑

These connectivity groups are set for network sources that represent a subway system and street network.

Note that metro entrances are the only source in both connectivity groups. Any path between the groups must travel through a shared junction. For example, a route solver may determine that a pedestrian's best route between two locations in a city is to walk to a metro entrance, board a subway train, take another train at a line-crossing station, exit at the third station, egress through the metro entrance, and walk to the final destination. Connectivity groups keep the two networks distinct, yet connect them at shared junctions.

Connecting edges within a connectivity group

Edges in the same connectivity group can be made to connect in one of two ways, set by the connectivity policy on the edge source.

If you set 'any vertex connectivity', line features are split into multiple edges at coincident vertices. Setting this policy is important if your street data is structured so that streets meet other streets at vertices.

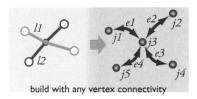

build with any vertex connectivity

Two polylines that geometrically coincide at a vertex can be split with a new junction if any vertex connectivity is set.

In this case, two polylines crossing at a shared vertex position will be split into four edges, with a junction at the vertex. Edges $e1$ and $e3$ in the network at the left are identified with the source feature class and object ID of line feature $l1$. Edges $e2$ and $e4$ are identified with the source feature class and object ID of line feature $l2$. Junction $j3$ will be a newly created system junction. Junctions $j1$, $j2$, $j4$, and $j5$ will either be system junctions as well or junctions from coincident points from a source feature class.

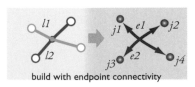

build with endpoint connectivity

The same two polylines will not split with a junction if endpoint connectivity is set.

If you set 'endpoint connectivity', then line features become edges joining only at coincident endpoints. In this case, line feature $l1$ becomes edge element $e1$ and line feature $l2$ becomes edge element $e2$. There will always be one edge element created for one line feature with this connectivity policy.

Building networks with endpoint connectivity is one way to model crossing objects such as bridges. Using elevation fields on planar data is another.

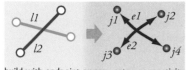

build with endpoint or any vertex connectivity

Crossing lines without coincident vertices or endpoints cannot produce junctions at the intersection.

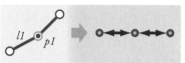

build with junction connectivity to override

A source point on a vertex of a source line with endpoint connectivity will form a junction in the network dataset if junction connectivity is set to 'override'.

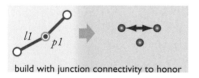

build with junction connectivity to honor

The same point and line will not form a junction if junction connectivity is set to 'honor.'

It's worth mentioning how crossing line features cannot produce connected edges. No connectivity setting will create a junction if they do not share any coincident endpoints or vertices. Street data for network datasets must be first cleaned so that either vertices or endpoints are present at all intended junctions.

If you need to remedy your street data, either use a geoprocessing tool to split crossing lines or establish a topology on these feature classes and edit street features while applying topology rules that enforce feature splits at intersections.

Once the sources are edited, you will need to ensure that all new features have shared vertices with intersecting features. This can be achieved by using the Integrate geoprocessing tool or by using topology.

Connecting edges through junctions across connectivity groups

You can only connect edges in different connectivity groups through a junction shared by both connectivity groups.

In the example of a multimodal system combining a subway system and street network, a subway entrance is added from a point source and is in both connectivity groups. The point location of the subway entrance must then be spatially coincident with the subway and street lines it joins.

When the point location for the subway system is added, whether it successfully becomes a junction depends on the junction connectivity policy. Like edges, junctions connect to edges at endpoints or vertices, depending on the target edge source's connectivity policy. However, you want to override this behavior in some situations.

Take a subway station as an example. The subway line connected to the station has an endpoint connectivity policy, but often you will want to place a station at an intermediate vertex. To do so, you will need to set a junction policy to override the default behavior connecting a junction to a given edge.

Setting the junction source's connectivity to override will override the default behavior of junctions forming at endpoints or vertices according to the edge source's connectivity policy. The default honors the edge connectivity policy.

Network attributes control traversability over the network. They specify properties of junctions, edges, and turns used as input for network solvers, such as time travel for road segments, speeds along a road, and whether a given road is one way.

Adding network attributes simultaneously adds them to the junction elements, edge elements, and turn elements. Inside the logical network, columns are added to the junction element, edge element, and turn element tables for the network attributes that you specify. Junctions, edges, and turns all share the same set of network attributes.

Since edge elements have two directions of travel, a set of network attributes is defined for both directions in the edge element table. Junctions and turns have one set of network attributes.

A network build creates network elements from source features which have network attributes. Network values can be assigned to network attributes through the use of evaluators, which are assigned to attributes of network sources. This is how attributes from source fields can be transferred to network attributes.

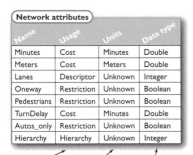

Network attributes

Name	Usage	Units	Data type
Minutes	Cost	Minutes	Double
Meters	Cost	Meters	Double
Lanes	Descriptor	Unknown	Integer
Oneway	Restriction	Unknown	Boolean
Pedestrians	Restriction	Unknown	Boolean
TurnDelay	Cost	Minutes	Double
Autos_only	Restriction	Unknown	Boolean
Hierarchy	Hierarchy	Unknown	Integer

Types of usage	Types of units	Types of data
Cost	Unknown	Boolean
Descriptor	Days	Double
Restriction	Hours	Float
Hierarchy	Minutes	Integer
	Seconds	
	Centimeters	
	Decimal degrees	
	Decimeters	
	Feet	
	Inches	
	Kilometers	
	Meters	
	Miles	
	Millimeters	
	Nautical miles	
	Yards	

Network attributes have a name, usage type, unit type, data type, and a Use By Default setting. If you set Use By Default to true, then other network attribute values will have default settings applied when you create a new network analysis layer.

Network attributes can also have zero to many parameters for modeling variable conditions, such as restrictions on vehicle height. See the ArcGIS online help for more information on parametrized network attributes.

Types of usage

The usage type for a network attribute specifies how you use it during analysis. These are the usage types.

- Costs measure attributes such as travel time or demand. Demand could be school children picked up by buses or snow to be shoveled. These attributes are apportioned along the distance along the edge, so that a travel of three minutes for a street line is reduced to 45 seconds if only 25% of that street is traversed by the solver. Network solvers use costs as impedances to find optimum routes through a network that minimize the accumulation of that impedance.

- Descriptors are additional characteristics that you can use in network analysis. Unlike costs, descriptors are not apportioned. The descriptor attribute applies to the entire edge element. Examples of descriptors are number of lanes and speed limits on a street, which can be used with evaluators to supply additional information for solvers to derive time of travel.

- Restrictions are used to identify elements that cannot be traversed or only in one direction. A one-way road would have a restriction set to True on the edge in the direction of prohibited travel. A pedestrian-only zone would have a restriction set to True on both edge directions.

- Hierarchy is the order or grade assigned to network elements. A street network has a hierarchy separating interstate highways from local roads. You can assign a preference (or deference) to parts of the hierarchy to find the shortest path. If set, hierarchies have two or three levels. Hierarchies aim to greatly reduce the subset of streets that a solver must consider to find the best route.

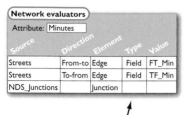

Types of evaluator
Constant
Field
Field expression
VBScript

Each network attribute has an evaluator. In this example, travel time minutes are calculated for both directions of a street.

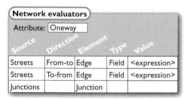

```
restricted = False
Select UCase ([Oneway])
  Case "N", "TF", "T": restricted = True
End Select
```

An example of a Field evaluator on a one way restriction attribute.

Evaluators

Evaluators assign values to network attributes. Evaluators are the mechanism to transfer attributes from source features to network elements. Evaluators can also perform a spectrum of logic, from simple constant assignments to complex scripts involving many fields. Each assignment has an associated evaluator that is responsible for returning attribute values for a given network element. These are the evaluators that assign values for the attributes of a source dataset.

The field evaluator is the most common way to assign values for a network attribute. It simply transfers the value of a specified field in the source feature to the network attribute. An example is transferring travel time attribute values from source features to network elements, which can then be used by the route solver. Note that one source feature may produce multiple edge elements; the field value for the one source feature gets transferred to all derived edges.

The field expression evaluator can take values from two or more fields in a source and combine them in a simple mathematical expression to derive a network attribute. For example, a source feature class may have information about distance and speed of travel, but not travel time. The field expression evaluator can take the distance value, divide by travel speed, and assign the calculated travel time to a network attribute.

The constant evaluator is the simplest of all evaluators and assigns a constant value, such as 0, to a network attribute of all network elements. This initializes a given attribute, such as setting all turn elements in a turn feature source to be restricted. Another example is setting travel time across junction elements to zero.

The function evaluator calculates attribute values by performing a multiplicative or logical function on another attribute value or parameter value. For numeric attribute types, the values are derived from an expression that multiplies the value of another attribute by some value.

The global turn delay evaluator assigns a cost value for transitioning between two edge elements based on the deflection angle between the two edges and the hierarchy attribute value of each edge. For example, it may take longer to make a left turn from a local road onto a secondary road than it does to go straight through an intersection on a secondary road.

The edge traffic evaluator reads historical traffic data tables and returns travel time values for an edge for a given time of day and day of the week. Only the most frequently traveled roads are present, so there are fallback attributes for returning travel times for roads not covered by the historical traffic data.

The VBScript evaluator lets you compose arbitrarily complex expressions that combine multiple network attribute values to calculate a new attribute. Unlike other evaluators, the VBScript is not run when the network is built, but when a network solver requires the result of that calculation. VBScript evaluators are designed for dynamic analysis of a network under changing conditions. Because this evaluator calculates its values at the time of analysis, it can affect the analysis performance.

TURN FEATURES AND ELEMENTS

Turns are essential for modeling transportation networks. An undirected network allows vehicles to travel freely in any direction. In reality, our street networks have many restrictions on left turns, right turns, and U-turns. Turns model these restrictions.

Turns are also useful for assigning additional travel costs on a maneuver. If making a turn between two major streets is constricted and slow, a turn can be added with that additional travel time.

Turn features

A turn is a maneuver from one network edge to another network edge through a sequence of connected intermediate edge elements. Turn features are line features with special turn attributes that overlay line features forming a maneuver along the transportation network.

Turns can have two or more edges. When you create a turn feature class, you define how many edges are allowed in a turn, which creates a multitude of attributes identifying the feature class, feature ID, and fractional position along that source line.

Turn elements

Turns can have network attributes that contain information about movement across the network. Examples of attributes are the time required for the maneuver and whether restrictions of vehicle class or pedestrian exist.

In a geodatabase, a turn feature class must be in the same feature dataset as the other network sources. With shapefiles, the turn shapefile must be in the same shapefile workspace (system folder) as the edge shapefile. Turn feature classes do not participate in connectivity groups (only junction and edge sources do).

User-defined fields in the turn feature class are used in the field evaluator of a network attribute in a similar fashion as fields from other feature class sources.

A turn from an on-ramp to a divided highway can capture any additional travel impedances, such as average merge signal wait.

Turns begin at a position expressed as a fractional position along the first source line feature and likewise end at a fractional position along the last source line feature. Positions are used when the line feature participates in any vertex connectivity or when the line feature has a multipart polyline geometry. This type of polyline produces multiple edges and junction elements, so the fractional position is necessary to identify which of the edges split from one multipart polyline and marks the beginning of the turn maneuver.

Simple feature class **Turns**				Geometry	Polyline		
				Contains M values	No		
				Contains Z values	No		
Field name	Data type	Allow nulls	Default value	Domain	Prec- ision	Scale	Length
OBJECTID	Object ID						
SHAPE	Geometry	Yes					
Edge 1End	String	Yes	N				1
Edge 1FCID	Long integer	No			0		
Edge 1FID	Long integer	No			0		
Edge 1Pos	Double	No			0	0	
Edge 2FCID	Long integer	No			0		
Edge 2FID	Long integer	No			0		
Edge 2Pos	Double	No			0	0	
Edge 3FCID	Long integer	No			0		
Edge 3FID	Long integer	No			0		
Edge 3Pos	Double	No			0	0	
AltID1	Double	Yes			0	0	
AltID2	Double	Yes			0	0	
AltID8	Double	Yes			0	0	

ELEVATION FIELDS FOR PLANAR LINES

In a network dataset, elevation fields model the logical stacking of roadways. Streets at ground level have an elevation value of zero, a bridge has an elevation of one, and a tunnel has an elevation of minus one. Elevation fields in a network dataset have no relation to terrain elevations.

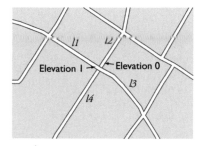

Overpass lines l1 and l3 have elevations of 1 at the ends adjacent to the overpass crossing and elevations of 0 at the other ends, where the roadway meets ground level. Underpass lines l2 and l4 have elevations of 0 at both ends. At the overpass junction, two system junctions are created: one with an elevation 0 and the other with an elevation 1.

Features with null elevation field values connect to other features with null elevation values, but do not connect to any features with numeric elevation field values. In addition, those features that have a numeric elevation field value at that site will still be honored.

This map shows the one-way flow direction and elevations inside a highway interchange. Despite the complexity of all the crossing roadways, four elevation values are sufficient to fully describe the connectivity of this intersection.

Most commercial street data contains planar line features and use elevation fields to separate traffic flows at bridges, overpasses, tunnels, and interchanges. For workflow efficiencies, street data is often kept planar. The network dataset uses elevations to refine the connectivity model so that traffic cannot turn from one roadway to another at different elevation levels.

Elevation is relative to ground level. The vast majority of edges and junctions have an elevation of 0, representing ground level. The raised ends of a simple bridge or overpass have an elevation of 1. A highway interchange typically has levels 0 through 3. A simple tunnel has an elevation of -1 at the lowered ends of the tunnel line.

When you create a network dataset, ArcGIS searches for any existing field that has a name indicating it is an elevation field. Some field names recognized from earlier network software are FNODE_ELEV and TNODE_ELEV, F_ELEV and T_ELEV, and F_ZLEV and T_ZLEV. You can also specify another field for elevation values. Elevation values must be correctly populated before you build a network dataset.

Elevation fields refine the connectivity where edge features join. They do not override the connectivity policies for edge sources but rather enhance network datasets to allow modeling of overpasses and tunnels. Each source within a network dataset does not need an elevation field for the network dataset to support elevation fields. Elevation fields are not added to sources that do not contain them.

Elevation fields respect connectivity groups; coincident endpoints of lines with the same elevation field value will only connect if they are within the same connectivity group.

Elevation fields are used if all features have a valid elevation value at a given location. Mid-span vertices on edge features do not have elevation values and are treated as having null elevation values.

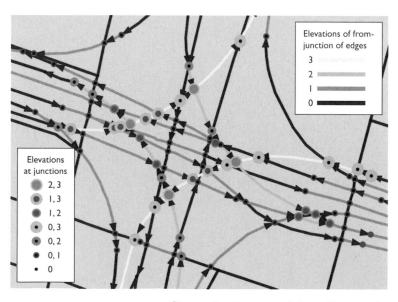

ASSIGNING HIERARCHIES

To understand hierarchy, think of how you use a road atlas to find a route. First, you locate your house and find a path through your neighborhood to the nearest main road in your town. From this main road, you will find the closest onramp to a highway in the general direction of your destination. Then you will find the offramp close to your destination and the final path there. Once you're on the highway, you naturally disregard the surrounding secondary street network. On a road map, hierarchies are clearly shown with the line symbols drawn for residential streets, major streets, and divided highways.

Solvers use hierarchy in network analysis to optimize path-finding performance. You can aggregate classification codes such as CFCC (Census Feature Classification Code) into three levels of hierarchy.

When using a hierarchy, the route solver first finds a path from a location on secondary road (low-level on the hierarchy) to the closest point on a feeder road such as a boulevard (mid-level on the hierarchy) to the closest on-ramp of a divided highway (high-level on the hierarchy), and then back down the hierarchy to the destination. Hierarchies free the solver from considering the whole network of secondary roads when finding a route.

Since hierarchies limit the streets considered for a route, it is possible that the solver might not find the shortest distance route. This is not a handicap for transportation systems—seeking shortcuts through neighborhood streets will almost always be slower than finding the most efficient way using available highways.

A route from points 1 to 2 at the right shows that paths through a street network typically begin at a neighborhood street, move to a major street, move onto a freeway, and then to a major street and neighborhood street. Hierarchy does not improve the accuracy of the route solution; it decreases route search time by an order of magnitude.

Network datasets contain an attribute model using restrictions, elevations, hierarchy, and turns to represent transportation network situations. You can combine these model elements in alternate ways to represent scenarios such as overpasses and intersections. Another influence on modeling is the type of source data and its properties. Typically, commercial street data has planar line features with elevations at endpoints and a classification used for hierarchy.

Transportation networks have long been modeled in GIS systems. Through commercial applications for logistics, best practices have been developed for modeling the parts of a road network. These are some common scenarios as properly modeled in a network dataset.

Street intersections

Traffic slows or stops in various ways at street intersections, and you can model how stops, lights, and slowing for turns affect route solver solutions. The global turn delay evaluator is an easy and powerful technique to model average wait times at left turns, right turns, and U-turns.

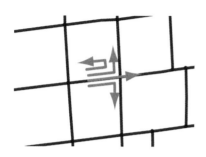

If you consider a simple intersection of two streets crossing, you will see that four turns are possible at each approach (left turn, straight, right turn, U-turn), and 16 turns are possible for the intersection.

Global turns are implicitly present at every transition between two adjacent edges within a network dataset where there is not already a turn feature present. The main purpose of global turns is to improve travel time estimates by penalizing turn movements not already represented or restricted by turn features.

A global turn delay evaluator works by, first, classifying two-edge transitions into four general turn types: left, right, reverse, and straight turns. (A reverse turn is like a U-turn, and a straight turn describes the movement of continuing directly ahead at an intersection.) Second, the evaluator penalizes transitions between edges based on the values you provide. For example, if you give global left turns a penalty of 15 seconds, the global turn delay evaluator finds all adjacent edges that are left turns and penalizes them by 15 seconds. The global turn delay evaluator is able to use hierarchies to classify and penalize turns more accurately. For instance, it can recognize a left turn from a local road to a secondary road and penalize it more than a left turn from a local road to another local road.

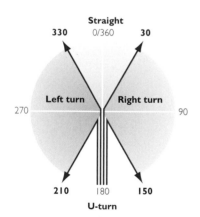

Classifying turns by deflection angles of 30, 150, 210, and 330 degrees is a common practice in the transportation industry. Dividing the range of turns into equal quadrants gives deflection angles of 45, 135, 225, and 315 degrees. Using global turn evaluators is an easy and effective way to assign travel costs throughout your network.

Global turns can be used with or without turn features present in a network dataset. Typically, turn features have more accurate penalties because each feature models a specific turn maneuver at an intersection; however, creating turn features for every turn, or even most turns, in a network is often infeasible because of the sheer quantity of turns that are possible. Alternatively, you can quickly set up global turns by defining turn angles for the four turn directions and assigning penalties for each turn type. You should note that global turns are typically not as accurate as turn features, because they are generalized and only penalize two-edge turns. Balancing these trade-offs between accuracy and ease of design can be accomplished by creating turn features at important intersections and invoking a global turn delay evaluator for the other areas of a network dataset.

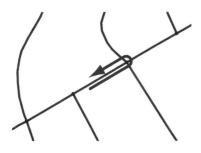

U-turn restrictions at an intersection can be modeled with a turn feature.

At major intersections, you can model maneuvers (paths through intersections) with turns if they need special wait times. You should not model stop signs and traffic lights with junctions because direction of travel is important; a junction for stops cannot specify how a two-way stop intersection is oriented.

U-turns at an intersection can be modeled with a turn feature. Remember that turns are not needed for routine, allowed maneuvers.

The majority of street intersections will not require turn features. Turn features will only be needed where turns are restricted such as no-left turn, and special transitions that may have a special cost, such as a highway ramp or left turn at a major intersection.

One-way streets

One-way streets should be modeled with a restriction attribute on edges in the from-to and to-from directions. In source features, one-way streets are typically coded with values of "TF," "FT," and " ". "TF" indicates that traffic flow is allowed in the to-from direction of edges; "FT" indicates that traffic flow is allowed only in the from-to direction of edges. This type of "FT" and "TF" coding is necessary in source features because the direction of digitization of a street usually has no relation to one-way restrictions. Once built in a network dataset, the edge elements have a one-way restriction attribute on each edge direction that replaced the to-from and from-to coding of street features.

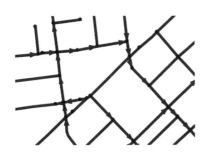

One-way restrictions not only limit traffic to one direction, they also restrict the available maneuvers at a junction. It is not necessary to place a turn when it is already forbidden by the one-way restrictions.

When a one-way street intersects another street, it is not necessary to place a turn to restrict a wrong-way maneuver onto the one-way street. The network solvers will use the one-way restrictions so that such turns are not necessary.

Divided roads

Divided roads can be modeled as a pair of parallel edges with one-way restrictions because they are two separate roadways that are physically separated by a median. On a divided road, a vehicle usually cannot make a left or U-turn in the middle of a road segment because of the physical barrier, so modeling divided roads with parallel edges is important in guaranteeing correct behavior for routing vehicles.

Intersections and turns are more complex at divided roads. Making a left turn on a divided road involves maneuvering through at least three edges instead of two edges encountered at a simple intersection.

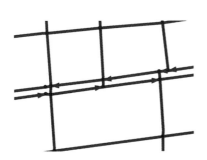

Modeling divided roads as parallel edges with one-way restrictions lets you finely model travel costs of turns.

More edges and junctions are required if divided roads are modeled with their separate roadways, but it simplifies the job of modeling turning maneuvers. Imagine an intersection of two divided roads. If modeled as two simple streets at an intersection, it's difficult to assign travel times. Attributes cannot be placed on edges that form a complex intersection. Modeling divided roads with parallel edges with one-way restrictions lets you control traffic flow and network attributes on the discrete parts of a complex intersection.

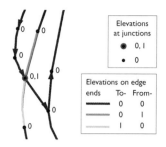

Elevations at junctions	
●	0, 1
·	0

Elevations on edge ends	To-	From-
———	0	0
———	0	1
———	1	0

This overpass is modeled with four planar edges that join at a crossing. The network build creates two system junctions at the crossing: one with an elevation of 0 and another with an elevation of 1.

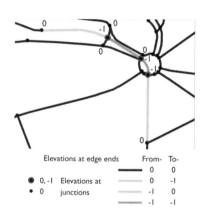

Elevations at edge ends	From-	To-
———	0	0
———	0	-1
———	-1	0
———	-1	-1

Elevations at junctions	
●	0, -1
·	0

A road entering a tunnel, underneath a roundabout on top of the hill

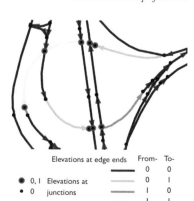

Elevations at edge ends	From-	To-
———	0	0
———	0	1
———	1	0
———	1	1

Elevations at junctions	
●	0, 1
·	0

A divided highway at an interchange

Overpasses

You can model overpasses in one of two ways, with nonplanar crossing edges or as planar edges with elevation fields.

If your street data is strictly planar, then elevation fields are required to distinguish roadways as they cross at different levels.

A simple overpass can be modeled with elevation fields with values set for the from and to endpoints of the edge. If an edge begins at ground level and moves to an elevated level, the from elevation attribute will be 0 and the to elevation attribute will be 1. If an edge represents traffic flowing under an overpass, the from and to attributes will be 0.

System junctions are created for each level at an intersection where streets cross over each other. In the case of a simple overpass, two junctions are created: the ground junction has an elevation value of 0 and the elevated junction has an elevation value of 1.

You can also use nonplanar crossing features to model overpasses. In this case, elevation fields are not required because roadways can be made to connect logically simply on their endpoint connectivity.

If vertices are coincident at the location of crossing, you must also set the edge connectivity policy for this source to be endpoint only.

Tunnels

Use negative elevations to model tunnels. In this example at left of a tunnel through a hill with a roundabout on top, the tunnel is separated from the traffic circle by elevation levels on the crossing edges and junctions.

In fact, this is modeled exactly the same as an overpass, except that it uses a negative elevation value. The junctions of a simple tunnel will have elevations of 0 and -1, depending on whether that edge is entering or leaving the tunnel. At each intersection point, two junctions will be present: one with an elevation of 0 and the other with an elevation of -1.

To suppress the drawing of a road inside a tunnel on a map, you can select for negative elevation values and exclude those feature with the layer definition query.

Highway interchanges

An intersection involving a divided highway can be quite complex, with elevated roadways crossing each other in three-dimensional space. This can be managed with a small set of elevation values, typically, 0 through 3 for an interchange of two divided highways.

As with overpasses and bridges, you can bypass the use of elevation fields by defining nonplanar edges that properly cross in space. This is an elegant solution and worth considering. These scenario discuss elevation fields because they are present in nearly all commercial street data.

An advantage of using elevation fields is that they can be used with finely defined layers in elevation groups to properly stack a complex highway overpass. If you model an interchange strictly with nonplanar crossing lines, you have no control over the drawing order of these elevated road features.

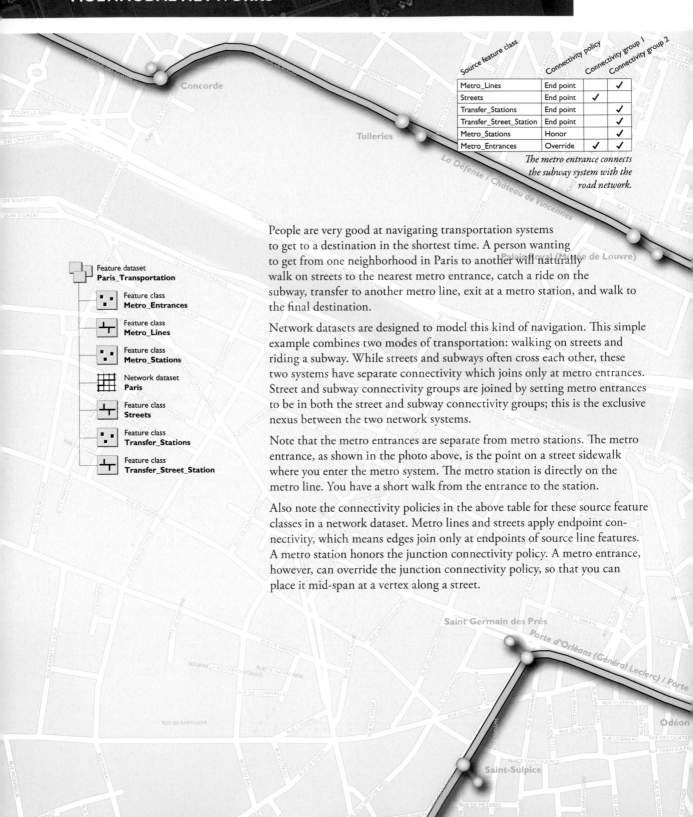

Source feature class	Connectivity policy	Connectivity group 1	Connectivity group 2
Metro_Lines	End point		✓
Streets	End point	✓	
Transfer_Stations	End point		✓
Transfer_Street_Station	End point		✓
Metro_Stations	Honor		✓
Metro_Entrances	Override	✓	✓

The metro entrance connects the subway system with the road network.

Feature dataset
Paris_Transportation

Feature class
Metro_Entrances

Feature class
Metro_Lines

Feature class
Metro_Stations

Network dataset
Paris

Feature class
Streets

Feature class
Transfer_Stations

Feature class
Transfer_Street_Station

People are very good at navigating transportation systems to get to a destination in the shortest time. A person wanting to get from one neighborhood in Paris to another will naturally walk on streets to the nearest metro entrance, catch a ride on the subway, transfer to another metro line, exit at a metro station, and walk to the final destination.

Network datasets are designed to model this kind of navigation. This simple example combines two modes of transportation: walking on streets and riding a subway. While streets and subways often cross each other, these two systems have separate connectivity which joins only at metro entrances. Street and subway connectivity groups are joined by setting metro entrances to be in both the street and subway connectivity groups; this is the exclusive nexus between the two network systems.

Note that the metro entrances are separate from metro stations. The metro entrance, as shown in the photo above, is the point on a street sidewalk where you enter the metro system. The metro station is directly on the metro line. You have a short walk from the entrance to the station.

Also note the connectivity policies in the above table for these source feature classes in a network dataset. Metro lines and streets apply endpoint connectivity, which means edges join only at endpoints of source line features. A metro station honors the junction connectivity policy. A metro entrance, however, can override the junction connectivity policy, so that you can place it mid-span at a vertex along a street.

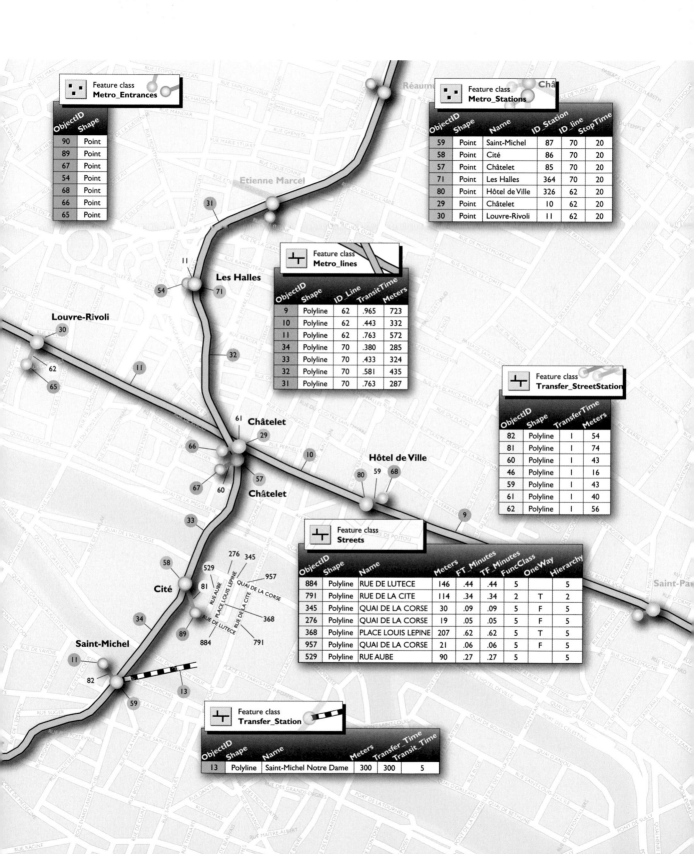

Feature class Metro_Entrances

ObjectID	Shape
90	Point
89	Point
67	Point
54	Point
68	Point
66	Point
65	Point

Feature class Metro_Stations

ObjectID	Shape	Name	ID_Station	ID_line	StopTime
59	Point	Saint-Michel	87	70	20
58	Point	Cité	86	70	20
57	Point	Châtelet	85	70	20
71	Point	Les Halles	364	70	20
80	Point	Hôtel de Ville	326	62	20
29	Point	Châtelet	10	62	20
30	Point	Louvre-Rivoli	11	62	20

Feature class Metro_lines

ObjectID	Shape	ID_Line	TransitTime	Meters
9	Polyline	62	.965	723
10	Polyline	62	.443	332
11	Polyline	62	.763	572
34	Polyline	70	.380	285
33	Polyline	70	.433	324
32	Polyline	70	.581	435
31	Polyline	70	.763	287

Feature class Transfer_StreetStation

ObjectID	Shape	TransferTime	Meters
82	Polyline	1	54
81	Polyline	1	74
60	Polyline	1	43
46	Polyline	1	16
59	Polyline	1	43
61	Polyline	1	40
62	Polyline	1	56

Feature class Streets

ObjectID	Shape	Name	Meters	FT_Minutes	TF_Minutes	FuncClass	OneWay	Hierarchy
884	Polyline	RUE DE LUTECE	146	.44	.44	5		5
791	Polyline	RUE DE LA CITE	114	.34	.34	2	T	2
345	Polyline	QUAI DE LA CORSE	30	.09	.09	5	F	5
276	Polyline	QUAI DE LA CORSE	19	.05	.05	5	F	5
368	Polyline	PLACE LOUIS LEPINE	207	.62	.62	5	T	5
957	Polyline	QUAI DE LA CORSE	21	.06	.06	5	F	5
529	Polyline	RUE AUBE	90	.27	.27	5		5

Feature class Transfer_Station

ObjectID	Shape	Name	Meters	Transfer_Time	Transit_Time
13	Polyline	Saint-Michel Notre Dame	300	300	5

After building your network dataset, you can perform several types of analysis. These are some of the key network solvers for a network dataset.

The route solver finds optimal paths through a network. The closest facility solver locates which facility, such as a hospital, is closest to an incident location. The service area solver determines an area served by a facility given an impedance, usually time or distance. Using the origin-destination cost matrix solver, you can interactively place stops, barriers, and facilities on the map, add them from addresses using a locator service or load them from a feature class. You can also load network objects and perform analysis through a geoprocessing script.

Finding the best route

You can find the best route along a set of stops that avoids barriers. The route solver takes as input a series of stops and barriers at points on features in the network dataset and applies these parameters:

- An impedance is the cost attribute to be minimized. A distance impedance finds the shortest route. A time impedance finds the quickest route.

- A restriction models a flow attribute of network edges, such as whether a street is one way. Other examples of restrictions are weight and height limits.

- A time window represents a time-of-day restriction on visiting a stop. Many stores limit delivery to hours with low traffic.

- Stops can either be traveled in the input order, or the stops can be reordered to further optimize the route.

- You can calculate a route, specifying whether U-turns are allowed everywhere, nowhere, or only at cul-de-sacs.

- Besides the impedance, you can select additional attributes to be accumulated and added to the properties of the route.

You can display and print driving directions after making your route.

Making analysis layers starts the network analysis. The route layer contains stops and barriers you input, with routes created by the solver.

The numerals in the stop symbols show the order they are to be visited.

When you don't place a stop or barrier directly on the network, they are not used in analysis and appear with lighter symbols and question marks.

Finding the closest facility

You can find the closest facility to an incident location that avoids barriers. Examples are directing police cars to a crime scene or finding the closest hospital to an accident. You specify these additional parameters when you apply the closest facility solver to an input set of facilities, incidents, and barriers:

- A cutoff value lets you restrict the maximum value of impedance, such as a limit of 15 minutes of travel time. This is useful for limiting the set of potential results when solving many incidents versus many facilities.

- You can find more than one facility. For example, set the number of facilities to three to find the three closest fire stations for dispatching trucks to a major fire.

- Attributes on the network including one-way restrictions and travel time impedances based on direction of travel may change routes and travel times when you travel in one direction and then the other. You can set whether the solver should calculate routes to or from facilities.

After finding the closest facility, you can display the best route to or from facilities, return the travel cost for each route and display directions to each facility.

Finding service areas

You can find service areas around locations that avoid barriers on a network. Service areas model area of accessibility along the network around facilities such as hospitals, retail stores, and fire stations. A network service area is a region that encompasses accessible streets within a specified impedance, such as a five-minute travel time or three-mile distance limit. In addition to impedance, U-turn policy, and restrictions, you can set these parameters:

- The default polygon break sets the extent of the service area, such as a five-minute drive time.

- Multiple polygon breaks can be set up to create concentric service areas, such as 2-, 3-, and 5-minute service areas.

- You can set polygon generation either to keep separate service areas around facilities or merge them by break.

- You can also exclude network sources from service areas. This is useful for multimodal networks, such as excluding rail lines for solving automobile accessibility.

Origin-destination cost matrix

You can generate an origin-destination cost matrix used for logistics planning if you are modeling distribution of goods, such as the flow of manufactured goods from a warehouse to store locations. You input a set of origin points (such as warehouses) and another set of destination points (such as stores) on a network dataset. A cost such as travel time can be calculated for the best routes between all possible origin/destination combinations.

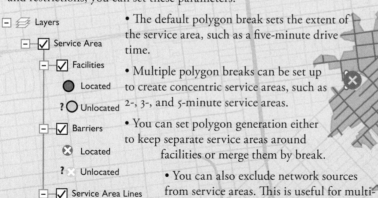

Travel time cost matrix
Warehouses

Stores	A	B	C
1	17.1	6.9	3.2
2	12.5	7.1	5.5
3	9.8	12.5	23.0

This type of matrix is identical to distance matrices you often see on road maps showing distances between major cities.

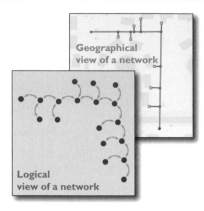

The geographic view is concerned with features and their geographic locations. The logical view considers only the connectivity of elements.

Just as with network datasets, there are two views of a geometric network: geographical and logical. The geometric network and its associated logical network represent these in a geodatabase.

The geometric network is the set of features that participates in a linear system. The geometric network corresponds to the geographical view of the network with point and line features that have positions and shapes. Examples of these are the ends of lines such as stream reaches and electric lines which join in space at points such as confluences and transformers.

The logical network is a set of junction and edge elements, without any geometry, which represent how the features are connected. The logical network is concerned only with the connectivity and traversibility of the system. You do not directly access or interact with the logical network; it is maintained by the geodatabase when the geometric network is edited. This abstraction of network features to a graph is what enables tracing analysis and sophisticated editing behavior such as rubberbanding of adjacent edges when moving a junction feature and maintaining connectivity when editing network features.

Features in a geometric network

A geometric network is a collection of features that comprises a connected system of edges and junctions. Features are connected based on geometric coincidence, if features exist at the same x,y coordinate; they are connected. Connectivity is not based on attribute values or relationships.

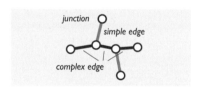

This simple geometric network has one complex edge, two simple edges, and six junctions.

There are three types of network features:

- Junction features represent the location at the endpoints of edges or where multiple edges are connected to each other.

- Simple edge features represent the line between two adjacent junctions, where resources enter one end of the edge and exit the other end.

- Complex edge features represent a set of connected lines with two or more junctions. Just like simple edges, complex edges allow resources to flow from one end to the other, but they also allow resources to be siphoned off along the edge without having to split the edge feature.

A network feature class is a collection of one of these types of network features: junction, simple edge, or complex edge. More than one network feature class can represent edges or junctions in a geometric network. A network feature class is associated with exactly one geometric network. Geometric network features cannot participate in other advanced datasets such as a topology.

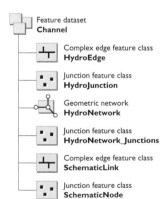

As shown in the catalog view, this geometric network has two complex edge feature classes and three junction feature classes.

Network features in a geometric network have all the same characteristics as other features. You can create as many feature classes as necessary for edges and junctions. You can add any attributes to these feature classes. You can define subtypes for major feature classifications and apply default values, attribute domains, and split/merge policies on attributes. You can establish relationships between network features and those not participating in the network. For advanced applications, you can extend a network feature class and create custom network features.

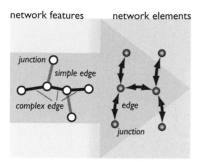

network features | network elements

Five edge elements are discovered from this simple network of one complex edge and two simple edges.

Simple and complex features must always have a junction at their endpoints; so if a junction does not exist when an edge is created, one is created automatically.

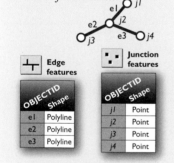

Geometric network

A geometric network contains the features that participate in a network. These features—junctions, simple edges, and complex edges—are stored in network feature classes.

Logical network

Junction connectivity table

JunctionID	Adjacent edges	Adjacent junctions
j1	e1	j2
j2	e1, e2, e3	j1, j3, j4
j3	e2	j2
j4	e3	j4

A logical network contains the connectivity of the network. The connectivity table lists all the adjacent junctions to a given junction, along with the edge that connects them.

Network features are edited within a framework that preserves connectivity and automatically updates network elements in the logical network. When a network feature is moved, its connected features are also moved. For example, when a junction is moved, the edges connected to the junction are moved along with it.

Elements in the logical network

Each geometric network has a logical network—a collection of tables in the geodatabase which store the feature connectivity, weights, and other information about the features in the geometric network. The logical network does not store coordinate values; instead, information is stored as individual elements for use in tracing and flow operations.

A junction feature is associated with one junction element in the logical network. A simple edge feature is associated with one edge element in the logical network. A complex edge feature is associated with a set of edge elements in the logical network, which is dependent on the connectivity of the complex edge feature.

Enabled and disabled state

All features in a geometric network have an enabled or disabled state. This is governed by a field and domain that is associated with each feature class when the geometric network is created. This state is used to control the ability for flow and tracing operations through specific features, while maintaining their connectivity. This is useful for modeling open switches or closed valves, where the state of the feature with respect to tracing or flow changes, but its location and connectivity do not change.

Network flow and weights

Sources and sinks are used in determining flow direction. Any junction feature class can take on the ancillary role of a source or a sink. A source is a junction from which a commodity flows. A sink is a junction where a commodity terminates.

Edges can have an indeterminate flow state. Edges that cannot be reached in the network are said to have an uninitialized flow. Edges have flow attribute with values of 'with', 'against', 'indeterminate', and 'uninitialized'.

Edges and junctions have any number of weights as an attribute. Weights store the cost of traversing across an edge or through a junction. Weights can be lengths, line capacity, electrical phase information, or slope.

Network analysis

A program that works with a geometric network to perform network analysis is called a solver. A solver takes as input a geometric network, with weights optionally specified, junction and edge flags specifying the origin of the trace, and junction and edge barriers specifying places in the network past which traces cannot continue. Solvers include upstream trace, downstream trace, isolation trace, and path trace.

A small portion of a natural gas utility system is shown as modeled with a geometric network. Features from five network feature classes combine in a geometric network and are related to elements in the logical network. This diagram shows a conceptual representation of how network features are modeled and have their connectivity stored in a logical network. Not all the details are shown, but the important tables of the logical network are illustrated here so that you can understand connectivity.

Junction feature class
Valve

OBJECTID	Shape	Material	Diameter	ValveID	Status
67	Point	PE	2"	2359	In Service

Junction feature class
Meter

OBJECTID	Shape	MeterID	Status
643	Point	2359	In Service
639	Point	2361	In Service
634	Point	2366	In Service
599	Point	2367	In Service
592	Point	2384	In Service
633	Point	2383	In Service
572	Point	2389	In Service
614	Point	2394	In Service

Valves and meters are modeled as junction features. Note that there are multiple junction feature classes in this geometric network. This allows modeling flexibility in representing distinct features such as valves and meters. One junction feature corresponds to one junction element in the logical network.

Junction feature class
Network_Junction

OBJECTID	Shape
221	Point
222	Point
223	Point
224	Point
225	Point
230	Point
231	Point
233	Point
234	Point

Creating a geometric network also creates junction features where an endpoint of a line is coincident with another line, but not occupied by another feature, such as meter or valve. These are called orphan junctions and placed in a new junction feature class. This is done so that the graph in the logical network is consistent and complete. These junctions are plentiful and not usually drawn on a map.

Feature dataset
GasUtility

Complex edge feature class
MainLine

Junction feature class
Meter

Geometric network
GasNetwork

Junction feature class
GasNetwork_Junctions

Simple edge feature class
ServiceLine

Junction feature class
Valve

Geometric network features

This is a portion of a geometric network, spanning gas utility main lines. The gas mains are modeled as complex edge features, one with six junctions and the other with five junctions. The secondary lines are simple edges which connect to the complex edge feature without splitting it.

Gas main lines are modeled with complex edge features to allow many service lines to connect without splitting the main line. A complex edge feature corresponds to one-to-many edge elements in the logical network.

Service lines are simple edge features because they require only two junctions: one that connects to the main line and another to the point of delivery, typically a meter. A simple edge feature corresponds to one edge element in the logical network.

Complex edge feature class
MainLine

ObjectID	Shape	Material	Diameter	Status	Shape_Length
1853	Polyline	Steel	2"	In Service	514.2
1622	Polyline	Steel	2"	In Service	347.9

Simple edge feature class
ServiceLine

ObjectID	Shape	Material	Diameter	Status	Shape_Length
376	Polyline	PE	1"	In Service	41.4
334	Polyline	PE	1"	In Service	34.5
395	Polyline	Steel	2"	In Service	20.3
354	Polyline	Steel	2"	In Service	42.1
361	Polyline	Steel	2"	In Service	56.0
401	Polyline	PE	1"	In Service	25.0
396	Polyline	PE	1"	In Service	22.8
405	Polyline	Steel	2"	In Service	17.8

The logical network

A geometric network is always associated with a logical network. Like a geometric network, a logical network is a collection of connected edges and junctions, which are continuously updated when editing network features. The key difference is that a logical network lacks coordinate values. Its main purpose is to store the connectivity information of a network along with certain attributes. A logical network is a graph stored in a set of tables in a geometric network, of which the three most important are shown here.

The centerpiece of a logical network is the connectivity table, which describes how network elements are connected. For every junction in the network, the connectivity table lists the adjacent junctions and edges—junctions at the other end of the connected edge. The geometric network maintains the integrity of the network through the connectivity table.

The logical network also contains a junction element table and an edge element table. The junction element and edge element tables provide a unique element ID that combines the feature class and the feature ID.

The logical network does not directly appear in ArcGIS applications. Rather, you interact with the geometric network. The logical network is the basis of the sophisticated behavior of the network features and enables tracing analysis.

Junction element table

Feature class	Feature ID	Element ID
Valve	67	1
Meter	643	2
Meter	639	3
Meter	634	4
Meter	599	5
Meter	592	6
Meter	633	7
Meter	572	8
Meter	614	9
Network_Junctions	221	10
Network_Junctions	222	11
Network_Junctions	223	12
Network_Junctions	224	13
Network_Junctions	225	14
Network_Junctions	230	15
Network_Junctions	231	16
Network_Junctions	233	17
Network_Junctions	234	18

Junction elements are stored in the junction element table. The logical network assigns an element ID for each feature class and feature ID combination.

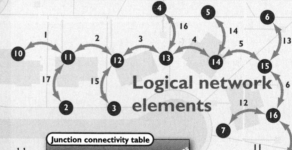

Logical network elements

Edge element table

Feature class	Feature ID	Element ID
MainLine	1853	1
MainLine	1853	2
MainLine	1853	3
MainLine	1853	4
MainLine	1853	5
MainLine	1622	6
MainLine	1622	7
MainLine	1622	8
MainLine	1622	9
ServiceLine	376	10
ServiceLine	334	11
ServiceLine	395	12
ServiceLine	354	13
ServiceLine	361	14
ServiceLine	401	15
ServiceLine	396	16
ServiceLine	405	17

The junction connectivity table establishes how maintaining a list of adjacent edges and junctions for each junction connects junctions and edges.

The edge element table stores edge elements. The logical network assigns an element ID for each feature class and feature ID combination.

Note that the two gas mains have multiple edges in the edge element table. This is because they have six and five junctions, respectively.

Junction connectivity table

JunctionID	Adjacent edges	Adjacent junctions
1	9	18
2	17	11
3	15	12
4	16	13
5	14	14
6	13	15
7	12	16
8	11	17
9	10	18
10	1	11
11	1, 17, 2	10, 2, 12
12	2, 15, 3	11, 3, 13
13	3, 4, 16	12, 14, 4
14	4, 5, 14	13, 15, 5
15	5, 6, 13	14, 16, 6
16	12, 7, 6	7, 17, 15
17	11, 8, 7	8, 18, 16
18	10, 9, 8	9, 1, 17

HOW CONNECTIVITY IS ESTABLISHED

Convert existing data

Import into new or existing simple feature classes

Build a geometric network from simple feature classes

Specify feature class roles

Identify use of flow direction and weights

Establish connectivity rules

Edit geometric network

You create a geometric network by specifying which line and point simple feature classes you want to participate in your geometric network, specifying into which role each feature class should belong and identifying if flow direction or weights are needed. Connectivity rules are set once the network has been created.

Clean data is the key to making a geometric network. Prevent undershoots or overshoots where line ends meet, either at a vertex or an endpoint. Make sure where points fall on lines (to become junction features), they are coincident with either a line vertex or line endpoint.

If you are unsure of whether you can guarantee geometric coincidence among the points and lines of your data, you can guarantee geometric coincidence by specifying that data within a specified spatial tolerance be automatically snapped during the creation of your geometric network. While using this method will guarantee geometric coincidence the result may not be what you expect, due to the indiscriminate nature of the snapping process. Features will be snapped solely based on their x,y location; it is best to prototype the results of this method on subsets of your data.

You can also use geodatabase topology to ensure coincidence before creating your geometric network. A geodatabase topology is particularly useful because you can more easily identify and correct possible spatial coincidence issues on simple points and lines. This will allow the connectivity to be discovered correctly during geometric network creation.

Connectivity model for geometric networks

Understanding how connectivity is established for features is important when creating your geometric network or when creating or editing network features.

Connectivity model for simple edges

Connectivity for simple edges is established only at the ends of edge features. Mid-span connectivity will not be established, even if there is a vertex along the simple edge feature.

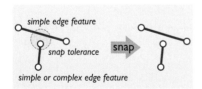

No connectivity is established.

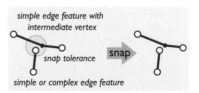

No connectivity is established. Mid-span connectivity on simple edge features is not established in snapping.

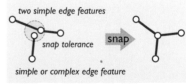

Connectivity is established. With simple edge features, only endpoint vertices are considered when establishing connectivity in snapping.

Connectivity model for complex edges

Connectivity for complex edges is established at the ends of features and at mid-span along the complex edge feature. A new vertex is created if there is no vertex along the complex edge where connectivity is established. Connectivity is not established between the mid-span of one edge and the mid-span of another edge; the end point of at least one of the complex edges must be used.

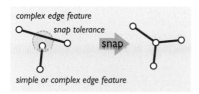

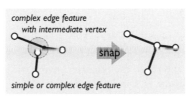

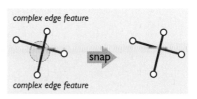

Connectivity is established. Intersection detection is performed along complex edges, and new vertices are inserted as required.

Connectivity is established. Mid-span connectivity on complex edge features is established in snapping.

No connectivity is established. Connectivity must be at an endpoint of one of the two edge features.

Connectivity model for coincident junctions

Junction features may only connect to each other through a simple or complex edge feature; connectivity between coincident junctions is not permissible. When coincident junctions are present which are geometrically coincident with at least one edge feature, the resulting connectivity will be nondeterministic. In other words, connectivity between any one coincident edge will only be to one of the coincident junctions.

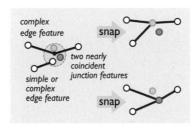

With two coincident junction features within a snap tolerance, edges will snap to one or another.

CONNECTIVITY RULES

In most networks, not all edges can connect to all other junctions. Also, not all edges can connect to all other edges through a specified junction. For example, a hydrant lateral in a water network can connect to a hydrant, but not to a service lateral. Similarly, a 10-inch transmission main can only connect to an 8-inch transmission main through a reducer.

Connectivity rules constrain the type of network features that may be connected to one another and the number of features of any particular type that can be connected to features of another type.

Connectivity rules let you easily maintain the integrity of the network features in a geometric network in an on-demand bases. While connectivity rules do not prevent features from connectivity, you can selectively validate features in the geometric network at any time, and generate reports as to which features in the network are violating one of the connectivity or other rules. The following are the connectivity rules for network features.

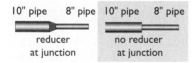

Yes, a service tap can be terminated with a meter.   *No, a meter cannot be connected to a main line.*

Edge–junction rule

You can specify which types of junctions are allowed to connect on an edge of a certain type. For example, electrical meters can only connect to low-voltage lines. Or, transmission towers carry only high-voltage lines. The edge–junction rule constrains which types of junctions can connect to a type of edge.

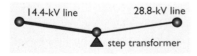

Yes, a valve connects exactly two lines.   *No, a valve cannot connect one or three or more lines.*

You can restrict the cardinality (count) of edges that connect at a junction. For example, a valve cannot be placed at the junctions of many lines; a valve restricts flow from one edge to another. You can limit the cardinality of edges to specified junction types. The edge–junction rule lets you precisely define the acceptable range of lines that can be connected at a junction.

Edge–edge rule

You can specify valid type combinations for connecting two edges at a junction. For example, a 10" pipe can only be connected to an 8" pipe through a proper reducer. This connectivity rule could be applied to two subtypes of a water pipe feature class for pipe sizes and a subtype of a junction feature class. The edge–edge rule establishes which combinations of edge types can connect through a given junction.

Yes, this reducer properly connects the two pipes.   *No, there is no reducer at the junction of two pipes.*

A side-effect of creating an edge–edge rule is the establishment of edge–junction rules representing the relationship between the junction and each of the two edges.

Default junction type

Both edge–edge and edge–junction connectivity rules can have default junctions associated with them. While default junctions are optional with edge–junction rules, they are required by edge–edge rules.

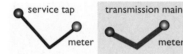

When you connect a 14.4-kV electric distribution line to a 28.8-kV line, a properly sized transformer can be automatically inserted.

When you connect one type of edge to another, you can specify a default junction type to be inserted. When a 14.4-kV line is added to an end–junction of a 28.8-kV line, a step-down transformer with the correct electrical ratings is assigned to the junction. The default junction type rule adds new junctions of that type at the free end of new edges added to the network.

There are times when you want to block or disable flow direction through a feature or the ability to trace through a feature in your geometric network. This may happen in an electrical network during a power outage, when overhead lines are downed by a storm. Since these power lines are no longer available, you would like to remove them from consideration during tracing operations.

Instead of deleting the feature or disconnecting the feature, you can disable the feature during tracing operations. A feature that is disabled acts as a barrier. When the network is traced, the trace will stop at any barriers it encounters in the network including disabled network features.

Disabled features do not participate in network flow; nothing flows into or out of the feature. Disabled features are useful for representing lines affected by open electrical switches or closed valves. The enabled/disabled state all affect how flow is established in a network.

The enabled or disabled state of a network feature is a property maintained by a combination of a field called Enabled which has a domain associated with it, the EnabledDomain. When adding new features to a network, they are enabled by default. The enabled state of a network feature is also stored within the logical network for quick access during solver operations. As with the connectivity of features, the enabled status within the logical network is maintained automatically.

Every junction and edge in a geometric network has an enabled or disabled state. In this example, edge e2 is disabled.

The enabled state of an element is also kept in the junction and edge element tables in the logical network.

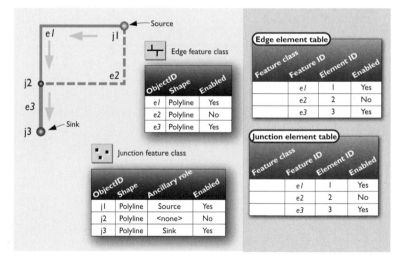

The network imposes flow direction by its configuration of sources and sinks, which may be represented by switches or outfalls, depending on the type of network. In utility network applications, the direction of commodity flow along edges must be an intrinsic part of the network.

If the geometric network is used for operational decision making, such as whether to close a switch or open a valve, you have to know if the decision will result in incorrect flow. In analysis, it is usually a requirement to know what features are downstream (with the flow) or upstream (against the flow) of some location.

A geometric network has a method to establish flow direction. This method decides how commodities flow in the network based on the current configuration of sources and sinks and the enabled state of each feature. The result of this method aligns the direction that commodities flow along each edge, either with the direction of the feature or against the direction of the feature, relative to its digitized direction.

The flow direction information is also stored within the logical network for quick access during solver operations. As with the connectivity of features, the flow direction status within the logical network is maintained automatically.

All line features have an implicit direction of digitization, which is the x,y coordinate order. In this simple network at right, flow goes opposite the digitized direction of edge e1, but with the digitized direction of e2 and e3.

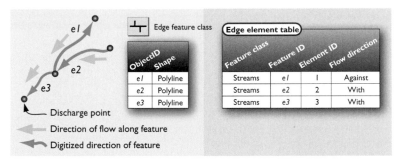

Sources and sinks

In a utility network, sources and sinks are used in determining flow direction. Any junction feature class can take on the ancillary role of a source or a sink. A source is a junction from which a commodity flows, such as a well-head pump. A sink is a junction where all commodity flow terminates, such as a wastewater treatment plant.

When creating a geometric network, you indicate whether or not a junction feature class contains features which can assume this ancillary role. If they can, the editor can specify whether an individual junction within the feature class is a source, sink, or neither.

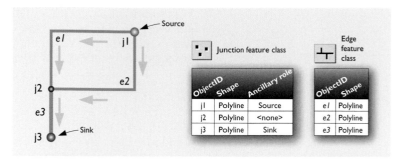

Junction features can have an ancillary role of source, sink, or neither. The role is stored in an attribute of the feature class, which is accessed by the establish flow direction method.

Indeterminate flow

It may not be possible to establish flow direction for an edge when the sources, sinks, and disabled features do not give enough information. An edge has indeterminate flow when flow direction cannot be established. Indeterminate flow occurs when the establish flow direction method cannot determine which direction commodities flow in a network.

It may not be possible to determine the direction of flow given a configuration of sources, sinks, and enabled features. This example is missing a source, so flow across the loop formed by e1 and e2 is indeterminate. The establish flow direction method will write "Indeterminate" as a flow direction when the flow direction cannot be established.

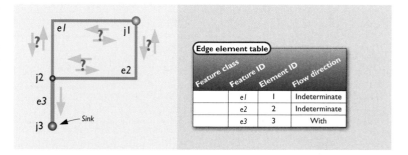

Uninitialized flow

Flow is said to be uninitialized when a flow is isolated because the edges are disconnected from the rest of the network (that has flow).

When establishing flow direction, edge features may be unreached because they are disconnected from the rest of the network. In this example, the unreached edges are disconnected because one of the junction features—a valve—is disabled.

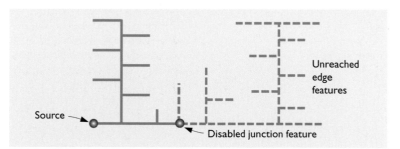

Most network applications need more than just simple connectivity. They also need to know the cost of traversing an edge or the cost of connecting two junctions. To support this ability, a geometric network can have a set of weights associated with it. For example, in a water network, a certain amount of pressure is lost when traveling the length of a transmission main due to surface friction within the pipe. Another example is the resistance to traversing an edge in an electrical transmission network, where the shortest path would be the path of least resistance.

When you build a network, you specify which attributes of edge and junction feature classes will become weights. Network weights are associated with one to many feature classes in your geometric network and are stored within the logical network. Weight values for each network element are derived from attributes on the corresponding feature. In the transmission main example above, the weight value is derived from the length attribute of the feature.

Edge and junction features can have any number of weights associated with them. The logical network stores weights.

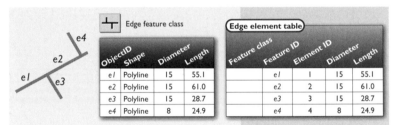

Each feature class in the network may have some, all, or none of these weights associated with its attributes. The weight for each feature is determined by an attribute for that feature. A network weight can be associated with only one attribute in a feature class, but can be associated with different attributes in multiple feature classes. For example, a weight called Diameter can be associated with the attribute Diameter in the water main feature class and with the attribute Pipe_dia in the water lateral feature class.

Junction features are associated with a single weight. For edge features, two weights can be used: one along the digitized direction of the edge feature (the from-to weight) and one against the digitized direction of the edge feature (the to-from weight). The digitized direction of an edge feature refers to the order in which the shape nodes of the feature are stored in the geodatabase. You can specify a different weight for each direction of an edge for cases where tracing an edge in one direction has a different cost from tracing it in the other direction.

A network weight value of zero is reserved and assigned to all orphan junctions. A network weight value of -1 indicates that the feature is impeded and cannot participate in tracing. Also, if a weight is not associated with any attributes of a feature class, then the weight values for all network elements corresponding to that feature class will be zero.

Network filters

You can use a weight filter to limit the set of network features that may be traced. A weight filter specifies which network features can be traced based on their weight values. A weight filter serves the same purpose as creating a selection of network elements based on a simple SQL query, except that the performance of the weight filter is much better.

Using a weight filter, you specify valid or invalid ranges of weight values for network features that may be traced. As with using weights to represent the cost of including a feature in trace results, a single weight is used for junction features and two weights may be used for edge features.

Bitgates

A bitgate is a special type of weight which provides an efficient way to represent categorical data and hierarchies. Most networks have no need for bitgate weights. Where bitgates are useful is when you want to capture categories of edges or junctions, such as the electric phase in a transmission line. In an electrical distribution network, phase (A, B, or C) can be thought of as a mode. The following example demonstrates this; the geometric network contains a 3-bit bitgate weight to model the three electric phases (A, B, and C). Each phase is represented by one of the three bits:

Phase A = 4 (third bit)

Phase B = 2 (second bit)

Phase C = 1 (first bit)

Each permissible numerical value must be entered in the weight filter range. For example:

- For tracing on Phase A (either state of the B and C phases is acceptable), the weight filter range will be '4, 5, 6, 7, 0.'

- For tracing on Phase AB (either state of the C phase is acceptable), the weight filter range will be '6, 7, 0.' Six represents the state of 'A AND B AND Not C,' while 7 represents the state of 'A AND B AND C.' Zero indicates that tracing is allowed on features that don't have an associated feature attribute for this bitgate weight; for example, orphan junctions.

In ArcGIS, network analysis is a procedure that navigates through the connectivity of the network to yield meaningful results such as finding all elements upstream of a point or the shortest path between two points. You can analyze networks in other ways, of course. For example, the basic selection tools found in ArcMap can select edge features and then calculate statistics about them, such as the total edge length by type of edge. This is certainly a valid analysis on a network, but it is not a "network analysis" because the network connectivity is not involved.

Solvers

An almost infinite variety of solvers exist for the many types of network analyses. ArcGIS provides a suite of solvers that address the more common types of problems. For less common types of network analyses, developers can create solvers using any programming language that can access the ArcGIS components.

A program that performs network analysis is called a solver, because it solves a problem, such as isolating flow to an edge by turning off a set of valves.

Inputs to the flow isolation solver in this example would be the logical network, the edge to isolate, and the set of junctions that are valves. The output would be the set of valves to turn off. The inputs and outputs of solvers have no rules, except that input always includes a logical network.

Solvers have user interfaces for specifying inputs and reporting outputs. Collections of solvers that perform similar tasks can usually be plugged into a common user interface framework. For example, the ArcGIS trace solvers are all accessed through a common toolbar. ArcMap is part of the user interface for a solver. Through ArcMap you graphically identify solver input such as start points for a trace.

Flags

A flag is a location on a network. Solvers use flags to represent a multitude of real-world objects, such as stops for a shortest path, start points for tracing, locations of valves, locations of services, and so on. Flags are not part of a logical network. They are used to describe any location in a network.

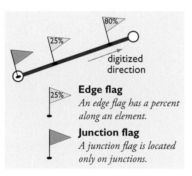

Edge flag
An edge flag has a percent along an element.

Junction flag
A junction flag is located only on junctions.

Flags are used to describe any location on a network. Examples include places to visit on a shortest path, the origin of a trace, a warehouse or service center, or a valve, switch, or transformer. Solvers rely heavily on NetFlags to describe input parameters.

Flags have two types: edge flags and junction flags. An edge flag can fall anywhere along the edge, from zero percent (from-junction) to 100 percent (to-junction), but junction flags must fall on a junction feature.

Barriers

Barriers are used by solvers to represent disabled logical network elements.

Barriers do the same job as setting an element's enabled/disabled state to disabled, except that barriers are not stored with the logical network—they are known only to the solver. Barriers are just a way to temporarily disable elements. Barriers are either on edge or junction elements.

You can use four methods to capture and represent barriers to a solver:

- Interactively add simple barriers.

- Use the features in your selection set.

- Disable feature classes.

- Apply a weight as a filter.

Tracing

Tracing means to follow the flow in a network until some condition is met. You are most certainly looking at using a trace solver to find the answer when you hear problems expressed as "search against the flow until you find a transformer," or "follow the flow upstream to the first discharge point," or "trace upstream and find all valves." The ArcGIS trace solvers include upstream trace, downstream trace, isolation trace, and path trace.

Weights

Choosing which edge or junction attributes should become weights in the logical network depends on your collection of solvers. Adding a weight to a network will not work if there are no solvers that can use it. For example, trace solvers typically do not use weights, but rather the connectivity information found in the logical network.

For example, suppose you have a water distribution network with a numeric attribute containing the pipe manufacturer ID. Adding this attribute is unnecessary unless you have a solver that can use it. Even with a shortest path solver, it would not make sense to find the shortest path based on manufacturer ID.

But suppose you had a solver that could return all junctions that share edges of certain characteristics. In this case, you may want to use this solver to find all junctions where pipes from one manufacturer connect with pipes from another. In this case, it might make sense to add manufacturer ID as a weight.

Below is a table of just a few possible weight attributes and the types of solvers that would use these weights.

Weight description	Used for
Length of edge	Shortest path solvers. Many solvers have a need for length.
Diameter of pipe	Solvers that calculate pressure or head in a network.
Impedance (electrical resistance)	Calculating voltage drop in an electrical network.
Time to traverse an edge	Shortest path solvers.
Number of lanes on a street	Calculating traffic capacity or congestion on a street.
Road classification	Describing network hierarchy in hierarchical shortest path solvers.
Miles per hour	A shortest path solver that allows dynamic calculation of weights.
Hazardous material route	A filter to find a path only on hazardous material routes.
Toll (cost to use a road)	Shortest path solvers based on monetary cost.

These are the network solvers that operate on geometric networks. Network solvers operate on the geometric network and honor the network flags, barriers, and flow direction set on the network.

In the illustrations below, the yellow elements represent the features on the geometric network that are selected by applying the solver in that situation.

Each of these solvers can also report the total cost of tracing all elements in the trace or the cost of a segment of the trace. For example, you can accumulate water flow on successive reaches of a stream network and calculate the flow contribution to a single stream reach.

Solver	Solver description	Solver application
Trace upstream sink flag *flow direction needed to solve*	Given an origin set by a flag, the *trace upstream solver* traces against flow, finding all upstream edges and junctions until stopped by barriers, sources, or the end of the network.	A water utility could use the trace upstream solver to determine valves to shut off water to a burst pipe. This solver can also be used to find sources of pollution from a monitoring station.
Trace downstream sink flag *flow direction needed to solve*	Starting from an origin set by a flag, the *trace downstream solver* traces with flow until stopped by barriers, sinks, or the end of the network.	This solver can be used to identify parts of a network affected by a resource flowing downstream, such as a chemical spill. You can also use this solver to calculate distances from sinks.
Find common ancestors source flags *flow direction needed to solve*	From each flag, the *find common ancestors solver* traces against flow to sources or barriers, then finds the features (ancestors) common to all traces.	When an outage occurs, an electric utility receives phone calls from affected customers. These locations can be entered as flags, from which this solver can narrow the list of the suspect transformers or downed lines that caused the outage.
Find loops *flow direction not required*	The *find loops solver* finds cycles or circuits in your network. Flow direction is not a consideration.	In certain utility applications, loops are considered faults to be identified. For example, any loop in an electric system is a short circuit and cannot be allowed. This tool enforces logical consistency of radial networks and can also be used to troubleshoot causes of indeterminate flow directions when loops are allowed, such as water utilities.
Find path *flow direction not required*	The *find path solver* discovers a path between two flags regardless of flow direction. If there is more than one path, only the first path found is returned.	This solver is used to inspect the logical consistency of a network and check for connectivity between two points.

Solver	Solver description	Solver application
Find connected *flow direction not required*	The *find connected solver* locates all elements that are connected to the edge on which the netflag resides. The connected elements are referred to as the connected component.	This solver can be used to identify the connected parts of the network and validate data integrity and connectivity.
Find upstream accumulation sink *flow direction needed to solve*	The *find upstream accumulation solver* traces on all elements upstream from origin (a netflag) and returns the total value of these elements.	This solver can generate flow statistics at a point, such as finding the number of facilities upstream from a monitoring station.
Find path upstream source *flow direction needed to solve* flag	The *find path upstream solver* determines a path from a netflag against flow to the source.	This solver can be used to assure logical consistency of the network and can be used in applications such as finding the source of pollution from a monitoring station
Find disconnected *flow direction not required*	The *find disconnected solver* is the inverse of the find connected solver; it finds all elements that cannot be reached from the netflag.	This solver can be used to identify isolated parts of the network so that they can be properly reconnected if required.

For gas and the water data, the most common network analysis is valve isolation tracing. A pipe is leaking or otherwise needs repair at a certain location, and the field operator needs to know which valves to shut off and which customers will lose service.

In this map, four valves will need to be closed and 129 customers will lose service while the repair is performed.

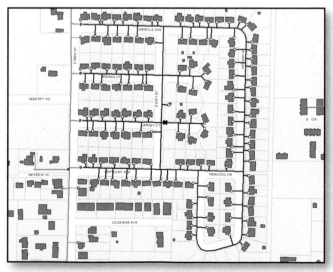

A network is a system of connected edges and junctions that transports people, manufactured goods, natural resources, energy, information, and communications.

An edge is bounded by two junctions. Examples of edge features are roads, river reaches, and utility lines. A resource flows on an edge from one junction to another.

A junction connects many edges. Examples of junction features are street intersections and stream confluences.

Networks have weights on edges which represent the cost or impedance of traveling part of the network.

Some networks have definite geographic edges and junctions such as street networks. Some networks have geographic junctions, but edge shapes are indeterminate or variable, such as an airline route system or animal migration patterns. Some networks have no geographic features at all but model relationships among people or objects such as networks of social acquaintances.

The essence of a network is its connectivity. Connectivity is rigorously described by a branch of mathematics called graph theory. Every network has its connectivity represented as a graph, a set of relationships between junctions and edges.

With GIS software, a graph is discovered by geometric coincidence of feature points and abstracted into a set of relationships persisted as a set of tables called the logical network.

The logical network does not store any coordinates or geometry; it is a set of junction and edge elements with their connectivity relationships and attributes.

The purpose of the logical network is to enable network analysis, which is done through network solvers, which implement algorithms from graph theory on the logical network to perform tasks such as optimized routing, allocation of resources, and finding closest facilities.

Real-world application includes two fundamental types of networks: networks with directed flow such as river networks, and networks with undirected flow such as street networks.

The geodatabase has two core network data models—the network dataset is optimized for undirected networks, especially transportation networks, and the geometric network is designed for directed networks such as utility applications.

Network datasets

A network dataset is a collection of edge, junction, and turn elements derived from network source feature classes. Network datasets are optimized for transportation applications.

Turns represent maneuvers through a connected set of edges and are used to prohibit or attach an additional cost to that maneuver.

Network source feature classes can be derived from two types of workspace: geodatabase and shapefile. Sources in a geodatabase offer the greatest modeling flexibility, shapefiles are easy to use and commonly available, and StreetMap data provides high data compression useful for large networks.

Junction elements are created from point features which are coincident with endpoints or vertices of source line features. Edge elements are created from source line features. Turn elements are created from source line features that represent maneuvers with turn attributes.

How connectivity is discovered is controlled by several properties of the network dataset: connectivity groups, edge connectivity policies, and junction connectivity policies.

Each edge source and junction source is assigned to a connectivity group. An edge source can belong only to one connectivity group. A junction source can belong to one or many connectivity groups. Features in separate connectivity groups can be traversed only through a junction common to both groups.

For edges within network sources in the same connectivity group, crossing line features can be split into multiple edge elements if they have coincident vertices and an edge connectivity policy of 'any vertex'. The other setting, 'endpoint', will not split any crossing lines.

Likewise, junctions can be discovered from source points if they are coincident with a vertex or endpoint. Further, you can honor or override this behavior for junctions and span multiple connectivity groups.

Network datasets have an attribute model with network attributes applied to all edge, junction, and turn elements. Network attributes can be assigned, calculated, taken from a source feature field, or dynamically calculated during network analysis.

Network datasets use elevation fields common on street data to model traffic flow on multilevel interchanges. Network datasets use hierarchy levels derived from a type classification to optimize solver performance in finding best routes.

For network analysis, stops and barriers can be dynamically placed on a network, with a solver applied on them. The analysis results are routes and service areas.

Network solvers on network datasets include finding the best route, finding the closest facility, finding service areas, and calculating origin-destination matrices.

Geometric networks

A geometric network is a collection of network features that comprise a connected system of edges and junctions. Geometric networks are optimized for facilities management applications.

The three types of network features are simple edge, complex edge, and junction. Complex edge features are used to model connected lines on which junctions can be placed on intermediate vertices. They are used to model primary lines that have many secondary lines connecting, such as meters to a gas main.

All network features in a geometric network reside in network feature classes in a feature dataset. Multiple network feature classes can participate as simple edges, complex edges, and junctions. A network feature class participates in one geometric network.

In a geometric network, sources and sinks are used to establish flow direction. These, together with disabled features, can be used to set

flow that is either 'with', 'against', 'indeterminate', and 'uninitialized' relative to the digitized direction of the edge.

Network feature classes have specialized behavior that updates their logical network elements whenever their features are edited. A geometric network is always in a topologically correct state.

Geometric networks support these types of connectivity rules.

- The edge–junction rule specifies which types of junctions can connect to a type of edge.

- The edge–edge rule establishes which combinations of edge types can connect through a given junction.

A geometric network is always associated with a logical network, which is a collection of connected edge elements and junction elements, without coordinates or geometry. The logical network enables powerful and efficient network analysis.

On a geometric network, solvers include downstream trace, upstream trace, isolation trace, and path trace. Weights represent cost attributes such as length, diameter, and electrical phase.

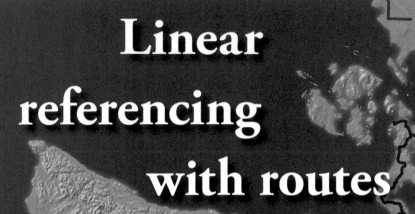

Linear referencing with routes

Linear systems such as roads, streams, pipelines, and railroads have many attributes that are defined at discrete locations or sections and these attributes often change with time. In the geodatabase, you can represent these complex systems with a simple model called linear referencing. This model is based on routes, which are line features with a defined measurement system. Linear referencing extends line feature classes for modeling events along routes and provides a framework for dynamically managing, displaying, and analyzing all the types of information that can occur at points and sections along routes.

5

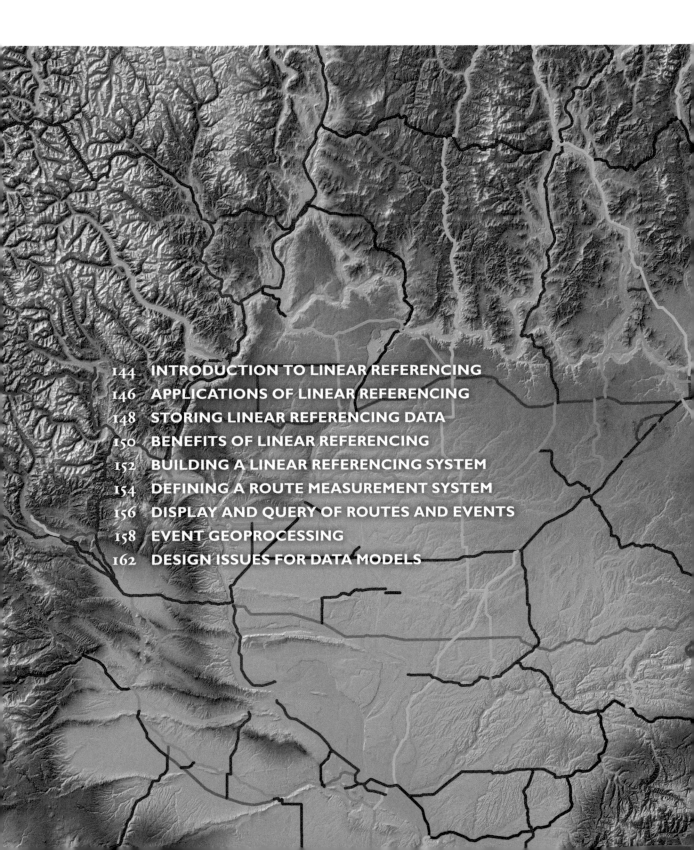

Milepost 12 along state highway 599 uniquely locates a position along that route.

For the public, highway routes are the most common example of a linear referencing system. Each highway has a unique designation such as 'Highway 599' and has a mileposting system that usually starts at zero miles at the beginning of the route.

Information about incidents and conditions along highway routes is typically collected at mileposts, which are linear measures along the route. The map shows a section of highway about 45 miles long. An agency that manages linear facilities will often prepare charts similar to this, showing information such as accident locations and pavement conditions along route measures.

Linear referencing is about locating features on routes. A route can be any linear feature such as a highway, river, pipeline, or GPS track. With linear referencing, features are located on routes by measurements made relative to a starting location instead of an x, y location. Measurements most often represent distances, but can also represent other values such as time.

Organizations such as highway departments, government agencies, and pipeline companies manage their asset inventory and incident data along linear networks by measurement systems such as mileposting. Examples are finding an accident at highway milepost 12.4, a navigational buoy at river mile 38, or a pumping station at 25 miles from the start of a pipeline.

Linear referencing provides two key benefits: it is a natural way to work with linear features and incidents that are recorded using measurements such as distance and time and it allows the modeling of many overlapping attributes on a route without splitting features.

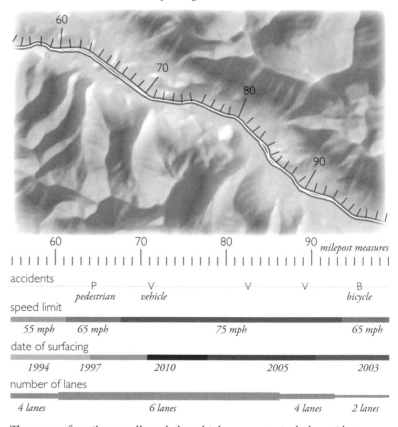

The types of attributes collected along highway routes include accidents, speed limits, date of surfacing, and number of lanes. Some of these attributes are located at discrete points, such as accident locations. These are called point events and are defined by a measure along a route. Other attributes are located along a section of a route, such as where a certain speed limit applies. These are called line events and defined by a starting point (from-measure) and ending point (to-measure) along a route.

Point and line events are stored in tables with their attributes. With linear referencing, you can place these features on a map and analyze their relationships on a route.

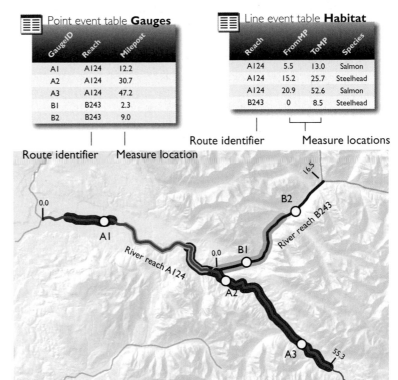

Point event table **Gauges**

GaugeID	Reach	Milepost
A1	A124	12.2
A2	A124	30.7
A3	A124	47.2
B1	B243	2.3
B2	B243	9.0

Line event table **Habitat**

Reach	FromMP	ToMP	Species
A124	5.5	13.0	Salmon
A124	15.2	25.7	Steelhead
A124	20.9	52.6	Salmon
B243	0	8.5	Steelhead

Route identifier   Measure location

Route identifier   Measure locations

Linear referencing is often used for mapping and managing river resources.

Sections of rivers are divided into river reaches, which begin and end at river confluences. Each reach is a route with a begin and end milepost value.

In this map, river gauges are mapped as point events and fish habitat ranges are mapped as line events. River gauges are shown with yellow point symbols. Steelhead fish habitat is shown with green lines, salmon fish habitat is shown with red lines.

Note that the steelhead and salmon habitat overlap along a distance interval. A key strength of linear referencing is the ability to model multiple and overlapping line features without duplicating feature geometries.

Point event table **Photos**

Waypoint	HikingTrail	Time	Photo
P12	NM-121	11:10	Raster
P13	NM-121	11:38	Raster
P14	NM-121	12:05	Raster

Route identifier   Measure location

In this map, measurements on a route are based on time instead of distance.

A GPS track is collected during a hike and calibrated with time values. The hike along a trail is a route with a start time measure of 10:50 and an end time measure of 12:17.

When a digital camera takes a photograph, the time is stored with the image metadata. You can build a point event table which includes the time the photo was taken. The photographs can be loaded as a raster field in the point event table.

This is an easy way to georeference photographs to the location they were taken. This technique is used for applications such as asset management by georeferencing images taken at a regular interval from a moving vehicle.

There are many applications of linear referencing. These are some examples.

Highways and streets

Agencies that manage highways and streets use linear referencing in a variety of ways in their day-to-day operations. These are some applications of linear referencing at highway agencies:

- Assessing pavement conditions

- Maintaining, managing, and valuing assets such as traffic signs and signals, guard rails, toll booths, and loop detectors

- Organizing bridge management information

- Reviewing and coordinating construction projects

Linear referencing also helps facilitate the creation of a common database that traffic planners, traffic engineers, and public works analysts can use for cross-disciplinary decision support.

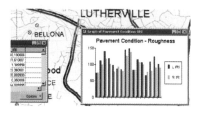

Analysis of pavement conditions on roadways. Positions for pavement conditions are stored with a start milepost and end milepost value.

Transit

Linear referencing is a key component in transit applications. Linear referencing facilitates these activities:

- Route planning and analysis

- Automatic vehicle location and tracking

- Bus stop and facility inventory

- Rail system facility management

- Track, power, communications, and signal maintenance

- Accident reporting and analysis

- Demographic analysis and route restructuring

- Ridership analysis and reporting

- Transportation planning and modeling

Map of a corridor study showing the number of traffic accidents along a stretch of highway. Accidents are recorded using highway milepost values.

Railways

Railways use linear referencing to manage key information for rail operations, maintenance, asset management, and decision support systems.

For example, linear referencing makes it possible to select a line and track and identify milepost locations for bridges and other obstructions that would prevent various types of freight movement along the route.

Further, dynamic segmentation can be used to display track characteristics as well as to view digital images of bridges and obstructions.

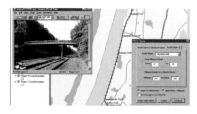

This display shows analysis of rail clearances along a rail line. Bridge inventory is recorded by distance along a rail line.

Oil and gas exploration

The petroleum industry manages tremendous volumes of data used in geophysical exploration. Seismic surveys, or shotpoint data, are used to help understand the underlying geology in an area.

The nature of seismic data is that it must be represented as both a linear object—the seismic line—and a collection of point objects—the shotpoint. Both the seismic line and the individual shotpoints have attributes, both must be maintained at the same time, and both are used in modeling applications. Linear referencing helps solve this problem.

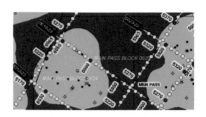

Map of seismic lines and shotpoints in the Gulf of Mexico. Shotpoints are collected at a distance measurement along a seismic line

Pipelines

In the pipeline industry, linear referencing is often referred to as stationing. Stationing allows any point along a pipeline to be uniquely identified. Stationing is useful for these activities:

- Collecting and storing information regarding pipeline facilities
- Inline and physical inspection histories
- Regulatory compliance information
- Risk assessment studies
- Work history events
- Geographic information, such as environmentally sensitive areas, political boundaries, right-of-way boundaries, and various types of crossings

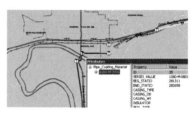

This map shows the examination of pipeline coating materials. Attributes of pipeline are recorded as a distance measurement called stationing.

Water resources

In hydrology applications, linear referencing is often called river addressing. River addressing allows objects, such as field monitoring stations, which collect information about water quality analysis, toxic release inventories, drinking water supplies, flow, and so on, to be located along a river or stream system.

Furthermore, the measurement scheme used in river addressing allows for the measurement of flow distance between any two points on a flow path.

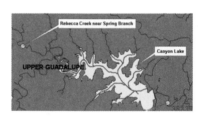

Monitoring stations along a hydrology network are identified. These positions are referenced by distances along each river reach. A reach is a section of a river between two significant confluences.

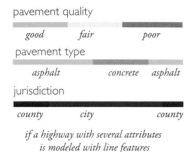

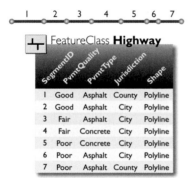

The problem with modeling multiple attributes on a linear system with line features is that you must create a new line feature wherever an attribute changes. In this simple example, seven line features are necessary for changes in three attributes. With additional attributes, the problem becomes worse, resulting in many small line features that become difficult to manage.

By separating attributes on routes from route features, linear referencing provides a solution to the problem of segmenting line features.

Linear referencing implements a form of database normalization which removes redundancy of attribute values and consolidates many small line features into one large line feature. This has considerable benefits for data management.

Organizations use linear referencing to locate and manage facilities along routes because it offers several advantages for geographic data modeling. The main benefit of linear referencing is the ability to associate multiple sets of attributes to a line feature without requiring underlying line features to be segmented each time that attribute values change. Another benefit is that linear referencing naturally models how personnel in the field locate themselves on routes.

The principle challenge of managing information along route systems is how to handle the many types of attributes that occur along a common set of lines. Along a highway route, a department of transportation manages diverse attribute data such as number of lanes, pavement material, speed limit, and roadway quality. A section of a route that represents a speed limit value rarely coincides with another segment of the same route modeling a certain pavement condition, roadway quality, or number of lanes.

If you were to use simple line features for these attributes on a route, then those features must be split whenever an attribute changes. But with the linear referencing model, the underlying line features do not have to be split when an attribute such as pavement quality changes.

Another benefit of linear referencing is that it follows a natural method for location along routes. It is easier for maintenance crews to locate themselves using odometers and signs rather than interpolating coordinates from a map. While GPS receivers now make it easier to find coordinate-based locations in the field, these organizations continue to use linear referencing because field personnel find it natural to use odometers and signage.

Yet another advantage of linear referencing is that it ensures that events are located precisely on routes. If x, y coordinates are provided to describe the location of incidents such as traffic accidents and those coordinates or street centerlines were not precisely defined, confusion can arise about which of two or more closely spaced roads contain a given incident.

Data modeling issues served by linear referencing

Linear referencing works well for organizations that manage linear systems because it handles four key data modeling issues not addressed well with simple feature classes:

1 Sometimes linear systems overlap, such as highways and bus routes. Linear referencing lets you build many routes from a common set of lines. This simplifies the data management of feature geometries.

2 Sometimes linear systems evolve over time, such as streets that change their name because a city's boundaries extend into a new area. Linear referencing enables you to perform overlay operations on linear facilities and update them as other features such as jurisdictions change.

3 Attributes of line systems overlap and usually have no correlation. For a highway, speed limits, date of surface, and number of lanes usually span different sections. With a line feature class, this would require segmentation into many small line features; linear referencing lets you manage route features and model attributes spanning overlapping sections.

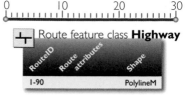

Route feature class **Highway**

RouteID	Route attributes	Shape
I-90		PolylineM

Line event table **PvmtQuality**

EventID	RouteID	FromM	ToM	Quality
I	I-90	0	6	good
2	I-90	6	20	fair
3	I-90	20	30	poor

Line event table **PvmtType**

EventID	RouteID	FromM	ToM	Quality
I	I-90	0	14	asphalt
2	I-90	14	26	concrete
3	I-90	26	30	asphalt

Line event table **Jurisdiction**

EventID	RouteID	FromM	ToM	Quality
I	I-90	0	4	county
2	I-90	4	28	city
3	I-90	28	30	county

With linear referencing, the segmented highway feature class example on the facing page is replaced by a route feature class with one route feature and three event tables, each with three events.

4 Attributes of linear systems evolve over time. An agency managing linear systems has a frequent need to compare conditions as they change over time. It is easier to update attribute values in normalized tables rather then segmented line features.

Linear referencing works by defining a measurement system on linear systems and separating route attributes into normalized tables associated with the routes by measure values.

Linear referencing

When you work with linear referencing data, there are two stages. The first stage is the process of creating your linear referencing data by building routes from line features and associating attributes, incidents, and entities on those routes through measure values.

Building routes solves the problem of segmentation; rather than create many small line features in a line feature class, a set of route features are created in a route feature class. Measurement systems can be defined on those routes by a process called calibration, which is discussed later in this chapter.

The second stage, once you've built your routes and events, is working with your linear referencing data. ArcGIS provides a set of linear referencing tools and commands for displaying and analyzing your linear referencing data. One of the most powerful capabilities is the ability to display your events as features in your map. This is achieved by a process in linear referencing called dynamic segmentation, which computes the map location (shape) of linearly referenced data (called events) on routes into points and lines on a map layer.

Anything associated with routes by measure location is called an event. Events can be attributes, incidents, or entities and are defined in event tables, which are simply tables that contain route identifiers specifying which route they apply to and measure values that are related to a route's measurement system.

The result of the dynamic segmentation process is a temporary feature class known as a route event source. A route event source can serve as the data source of a feature layer in ArcMap.

A temporary feature layer behaves mostly like any other feature layer. You can select a scale at which it should be visible, what features or subset of features to display, how to draw the features, whether to store it as a layer file, whether to export it, and so on.

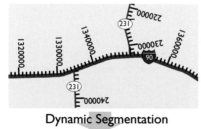

Dynamic Segmentation

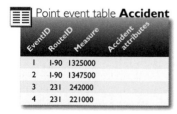

Point event table **Accident**

EventID	RouteID	Measure	Accident attributes
I	I-90	1325000	
2	I-90	1347500	
3	231	242000	
4	231	221000	

Route event source **Accident**

EventID	RouteID	Measure	Accident attributes	Shape
I	I-90	1325000		PointM
2	I-90	1347500		PointM
3	231	242000		PointM
4	231	221000		PointM

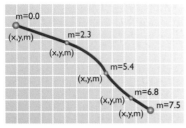

Route features have measure values for each vertex along the line in addition to (x, y) or (x, y, z) coordinate values.

Measure values are most commonly used for distances along routes, but other types of measure values can be used as well. One example is using travel time for measure values. An agency managing parks can use accumulated hiking times along a trail network to locate features of interest.

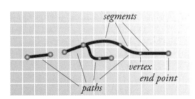

A single route feature has a polyline geometry which can have any number of paths, including disconnected and branching paths. Many routes have a single path with many connected segments.

A path can have one or more segments, which are either lines, circular arcs, elliptical arcs, or splines. Every vertex and end point in a route feature has either (x, y, m) coordinates or (x, y, z, m) coordinates.

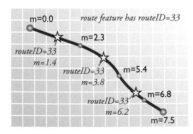

Events are located on a route feature by matching route identifiers and intepolating event measure values against the measure values at each vertex of a route feature.

In the geodatabase, features are stored in a feature class. Every feature has a geometry associated with it. This geometry is stored in a special field called 'shape.' There are three main types of feature: point, line, and polygon. The points and vertices of each feature geometry is located by two-dimensional (x, y) or three-dimensional (x, y, z) values.

Relative positioning with route measure values

Routes are a special type of line feature that have measure values, also known as m-values. M-values are used to designate relative positions along routes for linear referencing. In addition to (x, y) or (x, y, z) coordinate values that all features have, m-values are stored at each vertex of a route feature.

Measure values can represent any unit of measurement you choose, including miles, kilometers, meters, and feet for distances, or hours, minutes, and seconds for time intervals. Measure values are independent of the coordinate system of a feature class, however, it is not uncommon for the measure unit to be the same units as the line feature class' coordinate system.

Route features

Route features are stored in route feature classes. You create a route feature class by making a line feature class and specifying that it has measure values. Later in this chapter, you will learn how to assign measure values on routes in a route feature class.

A route feature is a line feature with a unique route identifier and measures defined at each vertex. A route identifier is stored in a user-defined field of either text or numeric type and can have any field name. Route identifiers are used to relate events like accidents and pavement conditions to route features.

The geometry of a route feature is based on the polyline geometry of line features with the addition of measure values. In ArcCatalog, you will see 'PolylineM' as the shape type. In most linear referencing applications, a route feature contains a single path with many connected segments, but any valid polyline geometry can be used for a route feature. The paths that make up a route feature may be disconnected or branching.

Events on routes

Point events, such as accident locations, and line events, such as pavement conditions along a section of highway, are stored in event tables.

Event tables contain rows with a unique route identifier, measure values, and other attributes of events. Like route feature classes, the route identifier in event tables is a user-defined field of text or numeric type. A route identifier field in an event table can have a different field name than the route identifier of a route feature classes, but their values, such as "Interstate 25", must match.

The route identifier for an event in an event table links that event to a route feature with a route identifier of the same value. The measure value for an event determines where that event is located along a route with a matching route identifier.

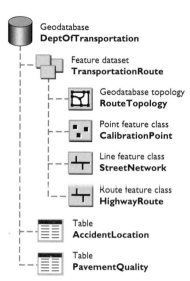

Point event tables have one measure value for the location of the event. Line event tables have a from-measure and to-measure representing the begin and end points of an event along a route. You can associate as many point and line event tables as you need to a route feature class.

There are three broad types of events that are commonly referenced to routes: entities, attributes, and incidents.

- Examples of entity events are stream gauges along a river network located using river mile and locations of traffic signs along highways recorded using miles from county boundary or highway interchanges.

- Examples of attribute events are road pavement conditions along highways and sewer conditions along segments measured by sensors.

- Examples of incident events are accident locations along a highway route or damage to a pipeline.

We use the term event to broadly describe anything that can be located along a route using measures.

Example of a route feature with events

The illustration below depicts one route with three line events and two point events and how they are organized into one route feature class and two event tables.

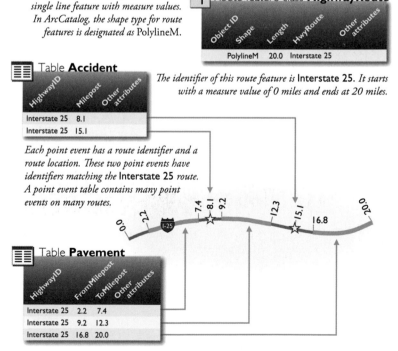

Each route in a route feature class is a single line feature with measure values. In ArcCatalog, the shape type for route features is designated as PolylineM.

*The identifier of this route feature is **Interstate 25**. It starts with a measure value of 0 miles and ends at 20 miles.*

*Each point event has a route identifier and a route location. These two point events have identifiers matching the **Interstate 25** route. A point event table contains many point events on many routes.*

*Each line event has a route identifier and from- and to- route locations. The three line events have identifiers matching the **Interstate 25** route. A line event table contains many line events on many routes.*

There are two basic approaches to building a linear referencing system comprising route feature classes and their associated event tables.

Building a linear referencing system from scratch

The first way to build a linear referencing system is to create an empty route feature class in ArcCatalog and use the standard edit tools in ArcMap to create new route features, assign route identifiers, and directly populate m-values at vertices of line features.

You would also create new event tables in ArcCatalog and add fields representing measure values and route identifiers. In ArcMap, you would create a new route layer by adding the route feature class and using the Routes tab to set the route identifier and add the associated event tables.

While this is a conceptually simple way to start, it's a laborious process and it's more likely that your organization already has measurement information and input features in some form.

Building a linear referencing system from existing data

The second way to build a linear referencing system is to start with an existing line feature class that contains the geometry of the routes you want to build and associate existing tables with data about the entities, attributes, and incidents you want to locate on routes.

For each set of line features sharing a common route identifier, such as a route ID or street name, one route feature is created in the route feature class. This is the more common scenario.

To build a linear referencing system from existing data, these are the basic design steps you will follow:

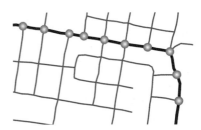

Routes are often generated from an input line feature class with many small line features that are usually segmented at intersections.

1   Identify the input line feature classes on which you want to locate events through measurements. The input line feature class may have many more line features than will be used in the routes you will define. For example, an input line feature class may have many types of roads from highways to dirt roads, but only the major roads might become routes in a route feature class.

2   Identify the event tables, in any format that ArcGIS can read, with entity, attribute, and incident data. Events are anything (entity, attribute, or incident) that can be located on a route, such as signs, pavement type, or accidents. These tables must have fields for a route identifier, measure values (a single measure value for point events or from- and to-measure values for line events), and attributes for the events. The fields for route identifiers and measures can have any name, but the field type must be consistent. These tables will become event tables associated with route feature classes.

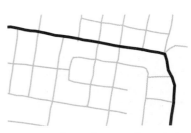

A generated route feature combines the line features that share a common route identifier into a single feature in a route feature class.

3   Identify the set of route feature classes required. If departments within your organization use different measurement systems on a common line network, you will need to create a route feature class for each unique measurement system. For example, a transportation agency will have

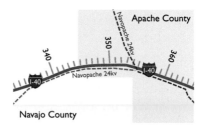

Locate features on routes

Line event table **Jurisdiction**

RouteID	FromMeas	FromMeas	Jurisdiction
I-40	331	345	Navajo County
I-40	345	366	Apache County

Line event table **Transmission**

RouteID	FromMeas	FromMeas	UtilityLine
I-40	335	355	Navopache 12kv
I-40	355	364	Navopache 24kv

Point event table **Sign**

RouteID	Measure	SignText
I-40	336	I-40
I-40	357	I-40

You can use existing point, line, and polygon features to create point and line event tables.

In this map, the I-40 freeway intersects Navajo and Apache counties. The intersection of those two polygons on the I-40 route creates two line events in the Jurisdiction line event table.

Two transmission lines partially follow the I-40 route. For the sections that they are close (determined by a distance tolerance), two line events are created in the Transmission line event table.

Two freeway signs are located on the I-40 freeway. They are within a distance tolerance and two point events are created in the Sign point event table.

multiple departments for asset management, pavement management, traffic analysis, and other aspects of transportation.

4 Identify how measure values will be assigned to each route feature class. One method is to calibrate routes using known measure values from calibration points, which have two essential attributes: a route identifier and the measure value at that point. Another method is to use route measure attributes (if they exist) from the reference line feature class. A third method is to use the geometric lengths of the input lines. This method presumes that the route measurement begins at zero at the start of the first input line.

5 Define the m- coordinate properties for each measurement system. For each route feature class, you will define a tolerance for the m-value, and the default m- tolerance value is usually suitable. Note that measure values are unitless, so the tolerance value for the map coordinates may be different from the tolerance value for the measure coordinates. The tolerance value is used to determine whether two measure values are considered to be identical.

6 Once your route feature class is built, you can use existing point, line, and polygon feature classes to build point and line event tables. For example, if you need a line event table for jurisdictions crossed by a route, you can create line events from the intersection of jurisdiction polygons with route features. You also locate line features along routes, such as utility lines next to roads, and convert them into a line event table. And you can locate point features, such as signs and valves, along routes and convert them into a point event table.

7 If you have more than one route feature class because of multiple measurement systems, it is best to place them in a common feature dataset. This will facilitate editing and allows you to manage the spatial integrity across multiple route feature classes with a topology. A topology is a collection of topology rules and is discussed in chapter 3, 'Vector Modeling with Features'.

8 Build a topology for your linear referencing system, especially if you are integrating multiple route feature classes along common reference line feature classes. A topology will enforce the spatial integrity of routes with a route feature class and routes across different route feature classes. This will assist you in editing routes which share geometry, such as the case when two highway routes overlap.

9 Test and refine your design using a file or personal geodatabase. Once you have tested your route feature classes and event tables with linear referencing functions in ArcGIS, then you are ready to migrate them to an ArcSDE geodatabase for enterprise-wide deployment.

These are the steps to create a linear referencing system. Once you have done so, then you are ready to use the functions in ArcGIS for display and analysis of routes and their associated events.

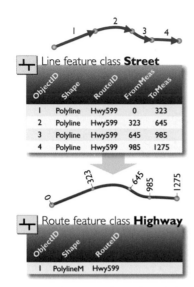

An input line feature class with two fields for from- and to- measures and a route identifier can be transformed into a route feature class with measure values assigned to the former endpoints.

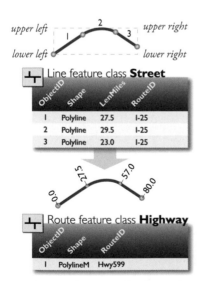

In this example, the input line feature class has a field called LenMiles containing line lengths in miles. Those values are accumulated into route measures. This example was built with a coordinate priority of "lower left". If you specified "lower right", the measure values would be reversed.

To perform linear referencing functions in ArcGIS, such as analyzing the spatial relationships among point and line events along a route, each route must have well-defined measure values. This means that measure values are properly assigned to each vertex along a route feature. Usually, measure values reflect distances, but they can also represent other values such as time intervals. Measure values usually increase or decrease progressively along a route, but routes can have measure values in any order.

When routes are created from an input line feature class, multiple line features sharing a common route identifier are combined to form a single route feature in the route feature class.

When you create a route feature class, you can use existing information in input feature classes to assign the measure values for each route. There are several methods to perform this and they are summarized below.

Calculating route measures from two field values

If your input line feature class has two fields with from- and to- measure values for the start- and end- points of input line features, you can apply a method to calculate route measures from this information. The input line feature class must also have a route identifier.

With this method, it is important that the line features are digitized in the direction of the measure values as the route measure values that you want to create. Normally, the to- measure of an input line matches the from- measure of the adjacent line in the route feature to be made, but this is not required.

There may be gaps between adjacent input lines. If there are, the measure values of the successive input line features control whether or not there is a gap in the measure values.

Calculating route measures from one field with a measure range

If your input line feature class has one field with a measure range, you can apply a method to calculate measures by accumulating measure ranges from the start to the end of the route. The measure range usually reflects a distance value, but can represent other values such as time intervals.

When you calculate measure values from a measure range field, you will need to provide more information about where the measures begin. By specifying a coordinate priority of upper left, upper right, lower left, or lower right, measure values will be calculated from the end point of the line closest to the selected corner of a rectangle formed by the set of lines.

If there are gaps between adjacent lines, you have two options for how the route measure values are calculated. You can either ignore the gaps in calculating measures between the nearest endpoints, or you can use the straight line distance of gaps in calculating measures.

Calculating route measures from geometric lengths

You can also apply a method to calculate measure values using the geometric lengths of the input line features. With this method, measure values reflect the map units of the input line feature class, such as meters or feet.

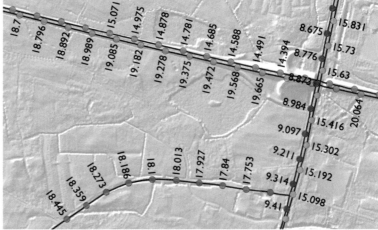

calculated measures that ignore gaps　　*calculated measures that include straight line distances for gaps*

As with the previous method, you will need to provide more information about where the measures begin by specifying a coordinate priority of upper left, upper right, lower left, or lower right.

If your input lines have gaps, you can choose to either ignore the gaps in calculating measures between the nearest endpoints, or you can use the straight line distance of gaps in calculating measures.

Using calibration points to define measures

A common way to define measure values on routes is to use calibration points. These points are usually collected in the field using distance measurement devices. Calibration points can be used in the initial definition of the measure values for routes and they can also be used to incrementally improve or update the measure values of routes.

Calibration points have two essential attributes: a route identifier and the measure value at that point.

This map shows calibration points color coded by route identifier and labeled with route measures collected with a distance measurement device. Many calibration points can be used to calibrate each route.

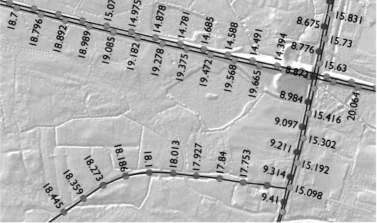

Calibration involves interpolating measure values on a route using calibration points within a tolerance distance of a route. The calibration points must have a route identifier that matches the identifier of the route within the tolerance distance.

There are two modes of calibration called interpolation and extrapolation and they specify whether measure values outside the range of calibration points are calculated.

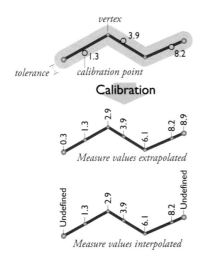

During the calibration process, a new vertex is created where each calibration point intersects the route with a specific tolerance, as shown in the illustration to the left. The measure value on these new vertices corresponds to the measure value stored as a point attribute.

You can specify whether vertices outside the range of calibration points will have their measure values calculated. This is called interpolation and extrapolation. When you calibrate with interpolation, measure values outside the range of calibrated points will be undefined. If measure values already exist at those outside vertices, then they will be unchanged. When you calibrate with extrapolation, then all vertices will have measure values calculated.

Either whole or partial routes can be calibrated. You can choose to interpolate between the input points, extrapolate before the input points, extrapolate after the input points, or use any combination of these three methods.

Like other geographic data, routes and events are displayed as layers in ArcMap. Route layers are similar to line feature layers but with additional methods for displaying routes.

Event layers have display methods in common with point and line feature layers, but event layers are special because they involve transforming route measures on point and line events in event tables (which are not feature classes) directly to the map.

Displaying routes

In addition to the standard display properties of lines, such as color and thickness, route layers have additional properties that you can set, such as hatch marks or symbols at intervals (defined in measure values) along the route. These hatch marks and symbols can also be combined with display text showing measure values.

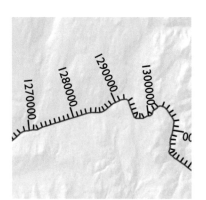

This map shows a route with two hatch intervals of 10,000 units and 1,000 units. The hatches for the 10,000 unit intervals are labeled.

Conceptually, hatch marks and labels are just like the divisions and labels on a ruler. You can define multiple levels of measure intervals to be displayed and you can selectively label levels of measure intervals.

In addition to displaying routes, you can perform spatial queries on a route. For example, you might want to find an accident location on a map using a route and milepost value. In ArcMap, you also have display tools that can locate the position of a route and a measure value or range on the map. The result will be displayed on the map. You can also point to any location along a route and find its measure value.

Finding and displaying route and measure errors

Sometimes, you might find errors in how measures have been defined along your routes. ArcGIS does not strictly enforce the validity of measure values along routes. For example, measure values usually increase or decrease along a route, but it is possible to unintentionally define routes with a mixture of increasing or decreasing values as well as undefined measure values.

In ArcMap, a route layer has display methods that show you where vertices along routes have no measure value (this is sometimes referred to as 'NaN', or 'not a number') and you can also find where measure values either do not increase or decrease progressively along a route. These are the most common issues encountered with verifying measure values.

You can correct measure values either by directly editing them in ArcMap or by adjusting and reapplying the calibration points with which you established the measure values for your routes.

You can also search for other route measure anomalies by using the definition query of the map layer and typing in an SQL clause for the invalid measure value condition.

Displaying point and line events

Recall that dynamic segmentation is the process of computing the map location and shape of point and line events stored in an event table. Linear referencing is what allows multiple sets of attributes to be associated with any portion of a linear feature.

The act of displaying point and line events on the map involves the calculation of map coordinates from route measurement systems. When point and line event tables are displayed on the map, they appear just as point and line features do.

Although point and line events are stored in tables, not feature classes, they can be displayed as a feature layer in ArcMap. This is done through the dynamic segmentation process and produces a temporary feature class known as a route event source. A route event source can serve as the data source of a feature layer in ArcMap.

For the most part, a feature layer based on a route event source behaves like any other feature layer. It is possible to decide whether to display it, the scale at which it should be visible, what features or subset of features to display, how to draw the features, whether to store it as a layer file, whether to export it, and so on.

You can edit a route event source in ArcMap. It is important to note that it is only possible to edit the attributes. The shapes of features from the route event source, however, cannot be edited because they are generated by the dynamic segmentation process.

Point event table **Sign**

EventID	RouteID	AtMeasure	Side	OffsetDist	Offset
1	Hwy599	12000	R	24	-24
2	Hwy599	16500	L	48	48
3	Hwy599	23000	R	36	-24

Point events such as signs can be displayed using a calculated angle and specified offset distances. From side and offset data, an offset field can be calculated with negative values specifying a right offset and positive values specifying a left offset.

Point event angles

When a point event is located along a route, it is often desirable to know the angle of the route where the event is placed. For example, you might need to rotate the marker symbol that is used to display the event so it is oriented to the route and not the map. Further, you might need to rotate a point event's label.

The dynamic segmentation process can calculate either the normal (perpendicular) or tangent angle. Further, it is possible to calculate the complement of these angles so that you can, for example, control the side of the route on which a rotated label appears.

An application of calculating point event angles is displaying signs along a highway. Although angles are calculated through dynamic segmentation, you will have to supply offset distances in your point event table.

Event offsets

In some applications, events with an offset are to be drawn to the right of the route. In other applications, events with an offset are to be drawn to the left. You can control which side of the route events with offsets are drawn; the default behavior is to draw negative offset values to the right side of the route in the direction of the measure values and positive offset values to the left side of the route in the direction of the measure values.

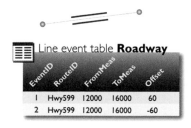

Line event table **Roadway**

EventID	RouteID	FromMeas	ToMeas	Offset
1	Hwy599	12000	16000	60
2	Hwy599	12000	16000	-60

Line events can also be drawn with an offset. You can draw a divided highway using offsets or you can stack line events from multiple line event tables using offsets.

You can use offsets for line events to simulate the display of a divided roadway. Line offsets can also be used for features offset to a roadway, such as guard rails.

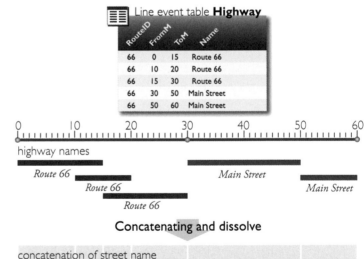

Given one or two event tables defined on a common route measurement system, you can apply event geoprocessing operators to produce an output event table.

The ability to perform route event analysis is a very powerful capability of linear referencing. These are the geoprocessing operations you can perform on event tables:

- You can summarize event data based on specific attributes by dissolving or concatenating (aggregating) event tables.

- You can find spatial relationships by overlaying existing event tables. There are two types of overlay operations: a union creates a new event table with every combination of input events and an intersection creates a new event table with overlapping events.

- You can transform route events from one measurement system to another. This is valuable when supporting multiple measurement systems or when updating event measure values to reflect the recalibration of a route.

Dissolving and concatenating events

Dissolving and concatenating events involves combining rows in line event tables and the results are writing to a new line event table. Concatenating will combine events in situations where the to-measure of one event matches the from-measure of the next event. Dissolving will combine line events when there is measure overlap.

Concatenating and dissolving events both combine adjacent rows in line event tables if they are on the same route and have the same value for the dissolve fields.

The purpose of dissolving and concatenating line events is to reduce the number of line events for better performance.

Line event table **Highway**

RouteID	FromM	ToM	Name
66	0	15	Route 66
66	10	20	Route 66
66	15	30	Route 66
66	30	50	Main Street
66	50	60	Main Street

Concatenating and dissolve

Line event table **Concatenate**

RouteID	FromM	ToM	Name
66	0	30	Route 66
66	10	20	Route 66
66	30	60	Main Street

Line event table **Dissolve**

RouteID	FromM	ToM	Name
66	0	30	Route 66
66	30	60	Main Street

Line-on-line overlay

An example of line-on-line overlay is to take an event table that describes pavement cracking and overlay it with pavement resurfacing dates. The results of such an overlay could be used to find the characteristics of the oldest paved sections.

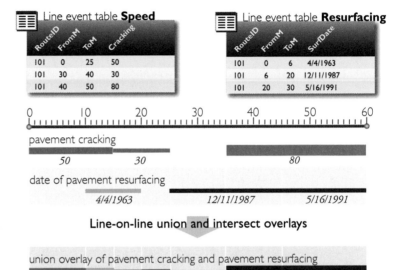

The union of two input line event tables is a line event table with each possible combination of overlay events. The intersection of an input line event table and an overlay line event table is a line event table with line events only where two line events overlap.

Line-on-line union and intersect overlays

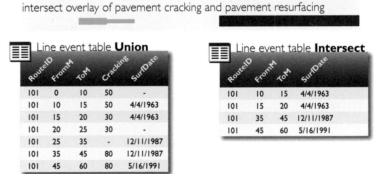

Line event table **Union**

RouteID	FromM	ToM	Cracking	SurfDate
101	0	10	50	-
101	10	15	50	4/4/1963
101	15	20	30	4/4/1963
101	20	25	30	-
101	25	35	-	12/11/1987
101	35	45	80	12/11/1987
101	45	60	80	5/16/1991

Line event table **Intersect**

RouteID	FromM	ToM	SurfDate
101	10	15	4/4/1963
101	15	20	4/4/1963
101	35	45	12/11/1987
101	45	60	5/16/1991

Line-on-point overlay

The intersection of a point and a line event table produces a point event table. The union of a point and line event table produces a line event table. An example of line-on-point overlay is analyzing highway characteristics where accidents occur, such as guardrail installations.

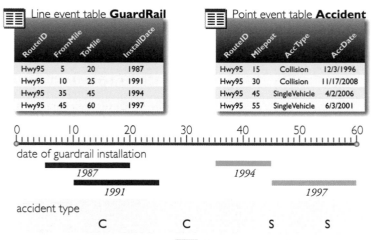

Line event table GuardRail

RouteID	FromMile	ToMile	InstallDate
Hwy95	5	20	1987
Hwy95	10	25	1991
Hwy95	35	45	1994
Hwy95	45	60	1997

Point event table Accident

RouteID	Milepost	AccType	AccDate
Hwy95	15	Collision	12/3/1996
Hwy95	30	Collision	11/17/2008
Hwy95	45	SingleVehicle	4/2/2006
Hwy95	55	SingleVehicle	6/3/2001

date of guardrail installation

accident type

Line-on-point union and intersect overlays

A line-on-point overlay involves the overlay of a point event table with a line event table to produce either a point or line event table.

In the union of a line event table and point event table, note that zero length line events are created where the point events occur.

union overlay of guardrail and accident

intersect overlay of guardrail and accident

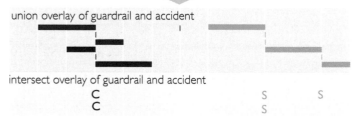
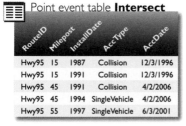
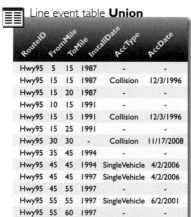

Line event table Union

RouteID	FromMile	ToMile	InstallDate	AccType	AccDate
Hwy95	5	15	1987	-	-
Hwy95	15	15	1987	Collision	12/3/1996
Hwy95	15	20	1987	-	-
Hwy95	10	15	1991	-	-
Hwy95	15	15	1991	Collision	12/3/1996
Hwy95	15	25	1991	-	-
Hwy95	30	30	-	Collision	11/17/2008
Hwy95	35	45	1994	-	-
Hwy95	45	45	1994	SingleVehicle	4/2/2006
Hwy95	45	45	1997	SingleVehicle	4/2/2006
Hwy95	45	55	1997	-	-
Hwy95	55	55	1997	SingleVehicle	6/2/2001
Hwy95	55	60	1997	-	-

Point event table Intersect

RouteID	Milepost	InstallDate	AccType	AccDate
Hwy95	15	1987	Collision	12/3/1996
Hwy95	15	1991	Collision	12/3/1996
Hwy95	45	1991	Collision	4/2/2006
Hwy95	45	1994	SingleVehicle	4/2/2006
Hwy95	55	1997	SingleVehicle	6/3/2001

Point-on-point overlay

The union of two point event tables is a point event table with all point events from both point tables. The intersection of two point event tables is a point event table with only coincident point events.

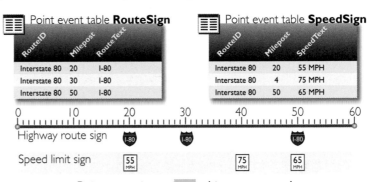

A point-on-point overlay involves the overlay of two point event tables to produce another point event table.

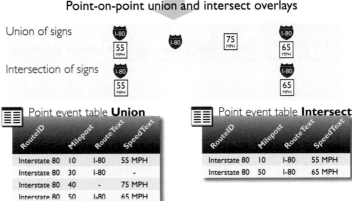

Transforming events

Sometimes it may become necessary to update the measure values in an event table. In many agencies, event data is collected against multiple route measurement systems. One agency might capture accident locations in one route measurement system while another might collect pavement condition using another route measurement system.

To combine events from different route measurement systems, you can use a linear referencing function to transform route events from one measurement system to another. This tool calculates the x,y coordinate location of each event in an input route reference and matches this to the corresponding measure values in a second route reference. The structure of the routes need not be the same. A new event table is written out with the events encoded in a different system of measures.

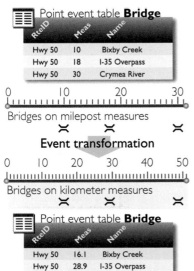

This example transforms bridge events from one measurement system to another.

This chapter presents an overview of linear referencing with the geodatabase. For more information, go to http://support.esri.com and search for the term 'linear referencing'. You will find many relevant technical papers, white papers, web-based help topics, data models, and discussion forum topics.

For further guidance on applying linear referencing, read Designing Geodatabases for Transportation *by J. Allison Butler. This book is published by ESRI Press and contains detailed data models and solutions to professionals responsible for modeling highways, rail systems, and navigable waters. All the design issues in this topic are discussed in detail in this book. In addition to linear referencing, this book provides guidance for applying geometric networks and network datasets for transportation agencies.*

Two topology rules that are applied to many route feature classes are:

1. Routes must be single part. For example, a single railroad route or pipeline route must logically be connected from start to end.

2. Routes cannot self-intersect. Most often, route features do not cross over each other. An exception to this restriction are delivery paths such as bus routes which can logically self-intersect.

When you drive on a highway and see two route identifiers mounted on a signpost, such as "Highway 285" and "Highway 84", then you are driving on a section of overlapping routes.

Linear referencing in the geodatabase offers a simple and flexible framework for modeling complex sets of events on route features. Some common data modeling issues arise at organizations that implement linear referencing. Several of the these issues can be summarized as:

- How to manage the spatial relationships among route features
- How to handle overlapping route sections
- How to handle the realignment of a section of a route
- How to handle multiple route systems with distinct measurement systems

Solving these data modeling issues depends considerably on an organization's business practices, so this topic offers some ideas and general guidance rather than standard solutions.

Handling spatial relationships among routes with a topology

In ArcGIS, topologies (collections of topology rules) are used to manage the spatial relationships between geographic features and to enable the sharing of geometry between features. For example, a topology can be used to ensure that multiple routes are spatially integrated on a single road network.

Topology rules allow you to model spatial relationships such as connectivity to ensure that all road lines are connected. Topology rules can also be used to manage the integrity of coincident geometry between different feature classes. Further, topology rules can be used to enforce the valid structure of individual linear features, such as routes should not be disjointed or routes should not self-intersect or self-overlap.

These are a few examples of topology rules that can be applied to route feature classes. You will specify topology rules that enforce the business practices of your organization.

Handling overlapping route sections

Many routes managed by transportation agencies, such as highways, commonly overlap. If an event has a route identifier and measure value, then how can you uniquely position that event on an overlapping section of two routes? There are several possibilities and they involve design trade-offs.

One solution is to use the lowest route identifier number in associated event tables. So "Highway 84" would be used instead of "Highway 285" to identify the route an event is associated with. This solution has the virtue of simplicity, but would require a gap in the "Highway 285" route.

Another solution commonly used is to create a new route feature for the overlapping section, which would have a combined route identifier such as "Highway 84 / Highway 285".

Yet another solution involves treating the overlapping section of routes as a line event. You would have many small route features wherever highway routes join with line events defining coincident routes. There is no single solution to modeling overlapping route sections—it involves how your organization defines routes and their measures and how event measure values are managed.

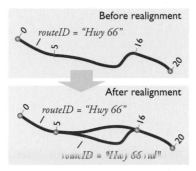

routeID = "Hwy 66"

0
1
5
16
20

After realignment

routeID = "Hwy 66"

0
1
5
16
20

routeID = "Hwy 66 rtd"

In this example, Highway 66 is realigned between milemarker 5 and 16. A new route feature is created from the retired section and the geometry of the existing route feature is updated with the new alignment. Measure values are recalibrated between milemarker 5 and 16 and measure values are not propagated outside the realignment section. You can either keep associated events on the retired section or migrate them to the realigned route.

This is a simplified example of a geodatabase with three route feature classes and distinct measurement systems. The three route feature classes are built from a set of anchor points and anchor lines which define the precise geometric shape of the underlying road network. Anchor points are usually located at street intersections and anchor lines are the spans between anchor points.

These are some example topology rules to manage the spatial relationships between anchor points and sections and multiple route feature classes:

1. Anchor sections must be single part.

2. Anchor sections must be connected.

3. The endpoints of an anchor section must be covered by an anchor point.

4. Each anchor point must be covered by the endpoint of an anchor section.

5. Routes in each route feature class must be single part.

6. Routes in each route feature class cannot self-intersect.

7. Routes in each route feature class must be coincident with an anchor section.

Handling the realignment of a section of a route

Another common data modeling issue is handling the realignment of a section of a route. This happens frequently with construction projects where a section of pipeline, railroad, or highway is improved and the alignment of the route changes.

The solution to this data modeling issue depends on whether you want to keep route measures outside the changed section to remain the same or be updated with new accumulated distance values.

If you want all route measures to be updated to the new accumulated distance values, then you can simply run a linear referencing function to recalculate route measures along that route.

More commonly, an organization will want to fix the measure values at the begin and end points of the realignment, so that measure values outside the realignment are not changed. If you do so, then how do you handle the associated events? An elegant approach is to create a new route feature for the retired section and modify the existing route with the new alignment geometry. Events that should be fixed to a permanent location, such as accident locations, are associated with the retired route. Events that should migrate to the new alignment, such as speed limit or pavement type, remain associated with the realigned route.

Handling multiple route systems with distinct measurement systems

Large organizations such as transportation agencies have multiple departments with the requirement to manage multiple route feature classes. Each route feature class often has distinct measurement systems on a common set of features, such as a street network.

Linear referencing in ArcGIS provides a complete framework for building this complex data model with the use of route feature classes, event tables, topology rules, and functions that can transform event locations from one measurement system to another.

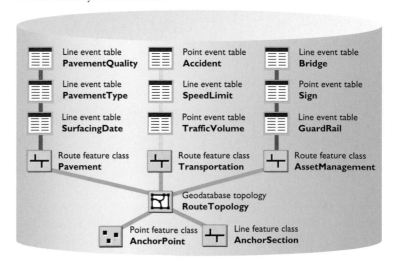

Finding places with locators

Locating a place by address is the most common geographic task that people engage in. Every organization has databases with tables containing addresses for customers and facilities. With ArcGIS, you can locate addresses on a map through an address locator in a geodatabase. An address locator contains reference data and methods for interpreting addresses. The reference data is most often a street network containing street names and address ranges for each street segment. The methods in an address locator encapsulate the addressing rules for a specific area or country. With an address locator, you can convert addresses to locations on a map and this process is called address matching or geocoding.

We can locate a place on a map in many ways.

We use grid lines on a topographic map (called a map graticule) to find a place based on its coordinate values such as latitude and longitude. We use grid reference values such as "H3" on a map gazetteer to locate a landmark such as a major building. We use route milepost values such as "mile 163" to find freeway exits on a highway map. But by far, the most common way that we locate places on a map is through addresses.

Many organizations manage database tables that contain addresses for customers or facilities. Every day, these organizations locate addresses on a map for purposes such as dispatching service or emergency vehicles, delivering packages and goods, analyzing customer locations, and allocating the placement of resources such as ambulances and fire stations.

Addresses are ubiquitous in information systems

Addresses are the most common form of geographic information and are an essential data source for many geographic information systems. Practically every residential and commercial location has an address.

Ideally, each address is unambiguously associated with a unique location, but in practice, addresses have several characteristics that present challenges for automating their placement on a map. These are some of the issues with finding locations for addresses:

- Addresses are entered into database tables by data entry personnel. Errors such as misspellings of street names or cities, incorrect postal codes, and incomplete addresses can and do occur.

- An address element may be represented in many ways. For example, a street type of "Avenue" might be spelled in full or abbreviated as "Ave.", "Ave", "Av", and other variants.

- Addresses follow rules that vary from one locale to another. One locale might follow a scheme where house numbers increase with odd-numbered addresses on the right side and even-numbered addresses on the left side. A nearby locale might define addresses based on street intersections.

- Addresses include street names that may change over time. Street names commonly change to commemorate important persons, such as local or national government leaders.

- A street may have a primary and multiple alternate names. When a highway goes through a town, "Route 66" might become "Main Street" and either street name might be used in an address.

- Some addresses may not have street names and can only be located within a broad area served by a post office. Examples are post office box and rural route addresses.

To work with addresses effectively in a geographic information system, ArcGIS implements a process called geocoding, also called address matching. With geocoding, you can enter one address interactively and find its location on a map or you can take a database table with millions of addresses and create a map layer with point features for every address found.

Besides addresses, a common way to reference a location is by a place name. You can enter place names such as "Bay Harbor Islands", "Biscayne Park" or "Miami" and an address locator will find that location for you.

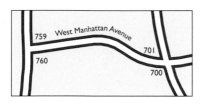

In the United States, addresses usually follow this convention: Odd house numbers are on the right side of the street (in ascending direction), and even house numbers are on the left side of the street.

Commercial and government street datasets are commonly segmented between intersections, and each street segment feature might have four fields for the 'Right-from', 'Right-to', 'Left-from', and 'Left-to' address value ranges.

House numbers can be geometrically interpolated between these values and placed with a small offset on the correct side of the road.

Steps for finding an address on a map

To understand geocoding in simple terms, it's useful to review how we naturally find address locations. Consider an example address:

759 West Manhattan Avenue, Santa Fe, NM 87505

To locate this address, we follow steps like the following:

1   Since this is an address in the United States, we presume it follows the most common national addressing convention, and we would adapt for local addressing conventions if necessary.

2   We would find a street map of the city of Santa Fe in the state of New Mexico.

3   Using a map gazetteer, we would find "West Manhattan Avenue" and take care to distinguish this street from "East Manhattan Avenue".

4   Address ranges for street segments commonly increment in values of 100 from a starting point in the city center. We would find the seventh street segment west of a central location in Santa Fe (the downtown plaza).

5   If our map shows address ranges for each block and this block has addresses ranging from 700 to 760, we would determine that the house number 759 is at the end of the block on the right side.

Geocoding with an address locator

Geocoding in ArcGIS follows the same logical steps as when we find an address on a map. All of the rules and geographic information necessary to find an address is stored in the geodatabase as an address locator.

An address locator encapsulates the rules for interpreting addresses (such as 'odd house numbers are on the right side of a street'), lists of standard street components and their abbreviations (such as 'Avenue', 'Ave', 'Boulevard', 'Blvd', and so on), plus the reference map data needed to find addresses.

Address locators are designed to handle the many possible variations of a valid address. If incorrect or insufficient information is provided in an address, you can inspect the suspect address matches and correct the addresses, refine the address locator rules, or improve the reference data until you get address matches that you are satisfied with.

The process of geocoding is done through an address locator. An address locator is stored in a geodatabase and contains rules for interpreting addresses as well as the reference data, usually a street network with street names and address ranges for each block. One or many addresses can be input into an address locator and the results are generated as points on a map.

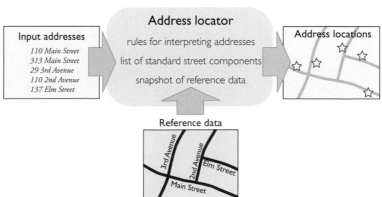

You can think of an address as a set of instructions to find a location. The address for ESRI corporate headquarters is

380 New York Street, Redlands, CA 92373

To find this location, you would first find the city of Redlands in the state of California. Within Redlands, you would find New York Street and interpret where and which side of the 300 block this address is located.

Locating addresses on a map is a process of narrowing your search. There are perhaps hundreds of roads named 'New York Street' throughout the United States and to find the right 'New York Street', you need to narrow your search to streets in Redlands, California. An address locator in ArcGIS follows a similar process.

The trouble with addresses is that they are subject to abbreviations and variations. Consider another address:

100 South King Street, Suite 500, Seattle, WA 98104

These are a few of the variations on that address that a postal delivery person could use to successfully deliver a letter:

100 S. King Street, Ste. 500, Seattle, WA 98104

100 S King St, #500, Seattle, Washington 98104

100 South King St., Suite 500, Seattle, WA 98104

A person can easily interpret any of these variations on an address, but computers need their input data to be precisely formatted. Address locators provide this precise formatting by breaking each address into its elements, recognizing standard components such as street types, and preparing a standard form of the address.

Address elements

Addresses are built from address elements. Some address elements common in the United States are identified in this address:

A house number specifies the building for an address and most house numbers fall within a sequential range along a street.

Directional information such as N, NW, W, SW, S, SE, E, or NE can be provided as either a prefix or suffix. This directional element is often necessary to distinguish between two or more otherwise identical addresses.

Streets have a name and a street type. Street types are frequently abbreviated. Some street types and their abbreviations are 'Boulevard' and 'Blvd'; 'Avenue' and 'Ave', 'Highway' and 'Hwy'; 'Street' and 'St'.

If an address references a building or set of buildings, such as a campus, then a sub-address is provided to locate a specific delivery point, such as suite, unit, apartment, or building.

Addresses have information about the city or region in which they are located. Postal services in each country use postal zones to enhance the automated handling of mail. In the United States, postal zones are five digit numeric identifiers called ZIP Codes. An example of a ZIP Code is '90210'. In the United Kingdom, postal codes are based on a hierarchical grid reference system consisting of letters and numbers. An example of postcode in the U.K. is 'HP21 7QG'.

When you use ArcGIS to locate a list of addresses on the map, the geocoding process identifies address elements and compares them against elements in the address locator's reference data. Address elements help in the geocoding search, pinpointing each address to a particular location.

Address formats

Addresses around the world follow a wide range of address formats. They share many of the same address elements, but have distinct address element types specific to a country, such as 'Camino' and 'Strasse', and these address elements may be arranged in different orders.

Within the United States, a common address format is a house number followed by a street name and type, perhaps a prefix direction or suffix, city and state, and ZIP Code. But some locales in the U.S. follow different address formats. For example, streets with Spanish names put the street type element first followed by street name, such as "Camino Real". Some cities use street intersections as part of an address; others use quadrant references such as NW. An address locator built for the U.S. must recognize these variations in address formats.

A typical address in Brazil is 'Rua Aurora 754, 01209001 Sao Paulo'. This address format is a near complete reversal of the address elements found in the U.S.—street type followed by street name followed by house number, then postal code followed by city. Brazil does not require an address element for the states. This information is encoded within the postal zone.

In German speaking countries, the street name and street type are commonly combined into a single name. For example, 'Wenderstrasse' combines the street name of 'Wender' with the street type of 'strasse'. The house number comes next, followed by the postal code and city name. As with Brazil, German addresses do not use address elements for the 16 states but provide regional location with the postal zone.

While address formats vary considerably around the world, every address consists of several address elements, presented in a particular address format recognized within that region.

Address locators are built to recognize the address formats and types of address elements that are particular to a region or country.

Typical address elements and address format in the United States

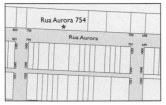

Typical address elements and address format in Brazil

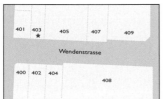

Typical address elements and address format in Germany

Address matching, or geocoding, is done in ArcGIS through a series of steps. You begin by preparing geographic data for matching addresses, called reference data. Then you determine which address locator style is compatible with your reference data and the format of the addresses to be matched. Next, you build your address locator with the selected address locator style and load feature classes with address information as the reference data. Finally, you use the address locator to perform address matching and rematch addresses which are not found in the first pass of address matching.

Build or obtain reference data

When you build or obtain feature classes for reference data, you may need to make modifications to the fields so that they coincide with the address elements required by an address locator style, such as house number, street name, and postal zone.

The first step to prepare for geocoding is to locate sources of geographic data for reference data. Reference data is loaded into an address locator and is a snapshot of geographic information. Reference data is built from feature classes with address information and can have point, line, or polygon geometry. Reference data should cover the geographic area for which you want to match addresses.

Select an address locator style

Address locators are designed to handle different types of input addresses. Most locators are designed for street addresses, but place names, cities, and postal zones can also be input addresses.

An address locator style is a template for building an address locator. You will select an address locator style that is compatible with the types of the addresses you want to match (street, place name, city, zone, or other). The address locator style must also be compatible with the fields representing address elements in your reference data.

Address locator styles prescribe a set of required and optional fields for address elements in the reference data. An example of a required field is street name. An example of an optional field might be prefix direction, representing an address element such as 'East'.

Build an address locator

You can build a composite address locator that combines multiple address locator styles for matching different types of input addresses or different sets of reference data.

An address locator is built as a dataset in a geodatabase. There are two basic steps for building an address locator: selecting an address locator style and loading reference data from feature classes. Reference data loaded into an address locator is a snapshot of geographic data. As new addresses are created for new developments over time, you can update those feature classes used to load the reference data and rebuild the address locator.

Perform address matching

Once you've built an address locator, you are ready to match addresses. You can either find individual addresses or load an input address table. When you match addresses, the result is one or many point locations on the map with the found location.

For addresses that do not match, you can rematch them by visually choosing among candidate locations or adjusting the threshold for match score values. Match scores are calculated to reflect the confidence level of a successful address match.

You will find matches for most input addresses, but there are a number of reasons why a match may be ambiguous or may fail. Some reasons include misspellings in input addresses or reference data, incomplete address elements (such as a missing prefix direction like 'West'), changes in street names over time, and new addresses not yet found in the reference data.

The process of geocoding uses sophisticated methods to break down and identify the components of an input address to match it with the address element information in the reference data. It's not necessary to understand the internal details, but the diagram below provides a summary of this process so that you can understand how an input address is matched to your reference data.

Successful address matching depends on the quality of the address information in the reference data. It's important that street names be spelled correctly, that necessary directional elements are provided, that correct house number ranges are provided for streets, and that cities and postal zone values are logically consistent.

An input address is parsed into address elements. The address locator searches for elements that correspond values such as prefix directions ('W') and street types ('Dr').

Some address elements are indexed.

A match score is calculated for candidate address matches.

A user-defined threshold value is applied to the match score and a list of candidate addresses are identified. An address is matched to the candidate with the highest score.

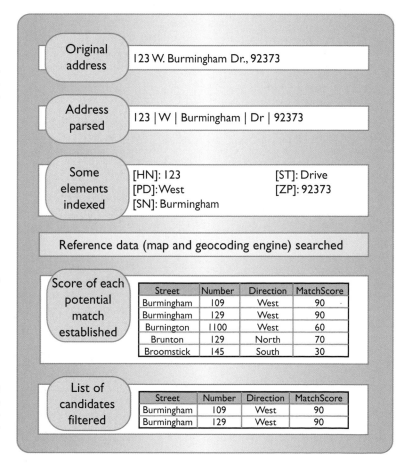

Original address

123 W. Burmingham Dr., 92373

Address parsed

123 | W | Burmingham | Dr | 92373

Some elements indexed

[HN]: 123   [ST]: Drive
[PD]: West   [ZP]: 92373
[SN]: Burmingham

Reference data (map and geocoding engine) searched

Score of each potential match established

Street	Number	Direction	MatchScore
Burmingham	109	West	90
Burmingham	129	West	90
Burnington	1100	West	60
Brunton	129	North	70
Broomstick	145	South	30

List of candidates filtered

Street	Number	Direction	MatchScore
Burmingham	109	West	90
Burmingham	129	West	90

An address locator is built by selecting an address locator style and loading reference data from point, line, or polygon feature classes.

You will build or acquire the reference data for your address locator from feature classes that contain address information. Your reference data must contain geometric information and fields for address elements sufficient to locate an address.

Considerations for building reference data

There are several factors for assessing the suitability of feature classes for building reference data in an address locator:

• The feature class for building reference data needs to span the geographic area in which you want to do address matching. If you want to geocode within a city, state, province, or country, your feature class needs to cover that area.

• The feature class for building reference data must have fields with address information to support the type of address matching you want to do. If you want to locate street addresses on a street network, the feature class should have fields for address elements such as house number, street name, city, and postal zone. If you want to locate landmarks, then a field with place names is necessary. If you want to locate a city or postal zone, then fields for those address elements should be in the feature class.

• The feature class for building reference data should have the desired resolution for the type of address matching you want to do. If you only need to locate places by city names or postal codes, then a feature class with city points or postal zone polygons will suffice. If you want to locate street addresses, you will need to judge whether interpolating a house number on a range of house numbers on street segments is adequate for your application. If you need more precise location of street addresses, you should consider using a feature class with points or polygons with unique address locations, such as buildings or parcels.

• The feature class for building reference data should be current and contain recently constructed roads or subdivisions. You can periodically update the reference data in an address locator with edited feature classes as you enter new streets and locations.

• The feature class for building reference data should have address elements properly entered. For example, a field with street names should have the names correctly spelled. Address elements such as street type and directions should be properly stored in the corresponding fields. The spelling of the elements should follow the locally recognized conventions.

The success of address matching depends on the quality of your reference data. When you start address matching, you will quickly learn how good your reference data is.

Building or acquiring reference data

The reference data for an address locator can be built from many sources.

- For the United States, you can use the StreetMap data and address locator that is available with ArcGIS. StreetMap comes with an address locator ready to use and you can consider whether this locator is appropriate for your use.

- You can purchase commercially available data from providers such as TeleAtlas® or NavTech™ or you can use government data such as the TIGER® database from the U.S. Department of the Census.

- If your organization is responsible for maintaining feature classes such as street centerlines, parcels, or buildings, and if these layers contain address information in some format, you can build reference data from these layers.

- You can also use reference data on the Internet served through ArcGIS Online services, such as StreetMap USA or Microsoft® Bing™ Maps.

The three map illustrations below show examples of reference data built from point, line, and polygon feature classes.

Polygon feature class

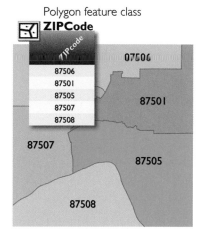

Reference data for postal zones requires only one field for the zone value.

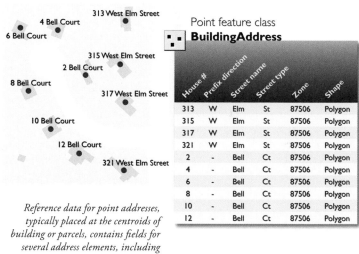

Reference data for point addresses, typically placed at the centroids of building or parcels, contains fields for several address elements, including house number, street name and type, and often city, state, and zone.

Point feature class **BuildingAddress**

House #	Prefix direction	Street name	Street type	Zone	Shape
313	W	Elm	St	87506	Polygon
315	W	Elm	St	87506	Polygon
317	W	Elm	St	87506	Polygon
321	W	Elm	St	87506	Polygon
2	-	Bell	Ct	87506	Polygon
4	-	Bell	Ct	87506	Polygon
6	-	Bell	Ct	87506	Polygon
8	-	Bell	Ct	87506	Polygon
10	-	Bell	Ct	87506	Polygon
12	-	Bell	Ct	87506	Polygon

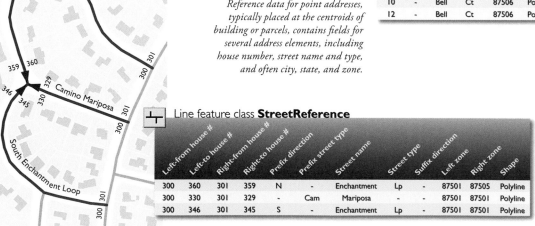

Line feature class **StreetReference**

Left-from house #	Left-to house #	Right-from house #	Right-to house #	Prefix direction	Prefix street type	Street name	Street type	Suffix direction	Left zone	Right zone	Shape
300	360	301	359	N	-	Enchantment	Lp	-	87501	87505	Polyline
300	330	301	329	-	Cam	Mariposa	-	-	87501	87501	Polyline
300	346	301	345	S	-	Enchantment	Lp	-	87501	87501	Polyline

In the United States, reference data for street centerlines usually has address ranges for the beginning and end of each street segment. Ranges are often provided for the left and right side of the street. Other fields model address elements such as street name, type, and postal zones.

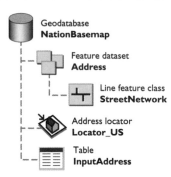

This catalog view of a geodatabase shows an address locator, a line feature class with street centerlines used to build the reference data in the address locator, and a table of addresses to be found on a map with the address locator.

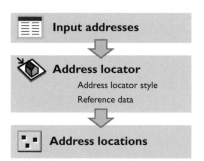

An address locator style is designed to handle the type of input addresses and the format of address information in the reference data. All address locators produce points on a map for the matched addresses.

You will probably find an existing address locator style that you can apply to your reference data, but if necessary, you can build a new address locator style.

Geocoding, or address matching, is done in ArcGIS with an address locator.

An address locator is a dataset in the geodatabase (or file folder) that contains information about local conventions for addresses (called an address locator style) and embedded map data such as street centerlines with address ranges (called reference data). When you perform geocoding, the address locator interprets an address using the address locator style and finds that address on a map using the reference data.

An address locator also contains information of how an address is parsed, searching methods for possible matches, and what output information of a match would be returned.

Address locator styles

When you build an address locator, you select an address locator style appropriate for the type of geometry (point, line, or polygon) of your reference data and the format of the address data you want to geocode.

The address locator style also defines the format of address fields required in the reference data. For example, if your reference data is built from street networks, an address locator for street addresses will usually require fields for address elements such as street name, street type, from- and to- address ranges for each street segment, postal zones, and city.

In the United States, two commonly used address locator styles are 'US Address—Dual Ranges' for reference data that contains street centerlines with attribute fields for left and right address ranges and 'US Address—ZIP 5-digit' for reference data consisting of polygons of ZIP Code boundaries. Address locator styles are built to handle the many variations of valid addresses in a country or region.

Composite address locator style

When you build an address locator, you also have the option to combine multiple address locators. The result is called a composite address locator.

For example, if you are building an address locator for the United States, most of your input addresses might be compatible with the 'US Streets—Dual Ranges', but some of your search data may not have full addresses. For example, if house numbers and streets are not given, you can define 'US Address—ZIP 5-digit' as a fallback locator to locate the general vicinity of a place. A composite address locator allows you to match input addresses with several variations in local address styles.

Common address locator styles

The illustration on the right shows six commonly used address locator styles for the United States. You can see how the address locator styles require certain address element fields in the input table and the reference data. An address locator style is also designed for the geometry of the reference data.

The address fields listed in the illustration describe the types of address elements. You can give the address fields in both the input table and the reference data any name you wish.

Common address locator styles for the United States	Address data in the input table	Reference data address fields	Map of reference data with sample matched address
U.S. Address—Dual Ranges *Finds input addresses with house number, street, and zone on reference data containing street centerline segments with from- and to-address ranges on the right and left side. This locator can place an address point on the correct side of the street. By default, addresses are placed at a 20-foot offset on the left or right side of the street.*	*All address elements are contained in a single field in the input table.* *Examples are* 320 Madison St. N2W1700 County Rd 105-30 Union St	House From Left House To Left House From Right House To Right Prefix Direction Prefix Type **Street Name** Street Type Suffix Direction **Left Zone** **Right Zone**	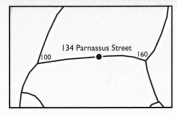
U.S. Address—One Range *Similar to the U.S. Streets and Zone address locator style, but works with street centerlines with from- and to-address ranges, without information about right and left sides. Addresses are located along the street centerline.*	*All address elements are contained in a single field in the input table.* *Examples are* 2 Summit Rd. N5200 County Rd PP 115-19 Post St.	House From House To Prefix Direction Prefix Type **Street Name** Street Type Suffix Direction **Zone**	
U.S. Address—Single House *Finds input addresses with street and zone fields on reference data where each feature represents one address. The reference data can be points, such as building or parcel centroids, or polygons, such as parcels and buildings.*	*All address elements are contained in a single field in the input table.* *Examples are* 71 Cherry Ln. W1700 Rock Rd. 38-76 Carson Rd.	House Number Prefix Direction Prefix Type **Street Name** Street Type Suffix Direction **Zone**	
U.S. Address—ZIP 5-Digit *Finds input addresses consisting of a 5-digit ZIP Code (postal code in the US). The reference data can contain points or polygons with ZIP Code values. If the reference data contains polygons, then the ZIP location is placed at the centroid of the ZIP Code polygon.*	*Five-digit ZIP Code values are contained in a single field in the input table.* *Examples are* 90210 87506 87112	ZIP Code	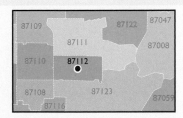
Gazetteer *Finds places such as mountains, bridges, rivers, and cities by feature names and geographic zone such as city, state, and country. The reference data represent places with either points or polygons and has fields for the place and geographical zone.*	*All place name elements are contained in a single field in the input table.* *Examples are* Leeds Castle, England Sapporo, Japan	Place City State Country	
Single Field *Finds input addresses consisting of a user-defined key field, usually a place name or landmark. The reference data can contain points or polygons with a key field with values that match the input address key field.*	*Features are identified by a text string, name, or code in a single field in the input table.* *Examples are* Cafe Cabrillo N1N115	Key field Note: A reference data field with light font style indicates that this field is optional.	

Often, one address locator can't do the job of matching all the addresses you need to geocode. For these situations, you need a composite address locator, which is a set of several address locators combined into one. These are several scenarios when the need for a composite address locator arises:

- You may need to match addresses that span a large region with different address styles, such as addresses from Mexico, the United States, and Canada. You will need a composite address locator to handle the different types and order of address elements that make up addresses for these countries.

- You may have multiple sets of reference data covering a region which cannot be merged because of differences in the address attributes. If you are matching addresses in the New England region of the United States, you might have distinct street datasets for the states of Vermont, New Hampshire, Massachusetts, and Maine. If these feature classes cannot be merged into one feature class, then you will need to build a composite locator to geocode in this region.

- You may have several sets of reference data which have varying degrees of completeness, geographic accuracy, and currency. For example, a municipality might have a set of address points that contain very accurate positions, but may not be complete. Another dataset may comprise street centerlines, which is complete, but not as accurate. Yet another dataset may contain parcels, including the latest subdivisions. A composite locator is very useful in this situation because you can search against your most accurate reference data first, and then fall back to other address locators for areas not covered by the first address locator or newly constructed subdivisions.

- You may wish to match varying types of address locations, spanning from detailed street addresses to landmark places, postal zones, and cities. You can build a composite address locator to search for a street address first, and then landmarks such as a hospital name, and then a postal zone if an address is incomplete.

Building a composite address locator

A composite address locator performs a multistage search against a number of address locators. The order of address locators in a composite address locator specifies which address locator is searched against first. Put your best address locator first in a composite address locator to get the best address matches.

To build a composite address locator, you begin with preexisting standard address locators. While in the initial phases of building your composite address locator, you should plan your search process. For example, you can search an address locator containing local road data first; then, if no satisfactory results appear, you can have the address searched by an address locator containing statewide or national roads. Finally, you may want to use an address locator that will search for a specific zone such as postal code or city.

The illustration on the facing page shows how a composite address locator is defined for the city of Santa Fe. This shows a common scenario for municipalities that have several types of available reference data of varying accuracy and completeness.

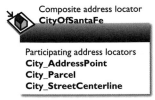

Composite address locator
CityOfSantaFe

Participating address locators
City_AddressPoint
City_Parcel
City_StreetCenterline

The city of Santa Fe uses a composite address locator with three participating address locators containing reference data built from three feature classes with progressive accuracy for address information.

The reference data for the City_AddressPoint locator is built from a point feature class containing points for each unique address. Address points are located at the centroid of the building or parcel. This is considered the best locator for address matching, but coverage for the entire metropolitan area is not complete.

If an address is not found in the City_AddressPoint locator, then the composite locator tries to find a match with the City_Parcel locator. Found addresses are located at the centroid of the parcel polygon.

If an address is not found in either the City_AddressPoint or City_ Parcel locators, then the composite locator tries to find a match with the City_StreetCenterline locator. Address locations interpolated on street centerlines are not quite as accurate as addresses found on address points or parcels, so this is the last of the locators used in the composite address locator. This locator is useful because it has full coverage of every street in the Santa Fe metropolitan area.

Local governments frequently match addresses using composite address locators. A composite address locator references several other address locators and chains them together. With a composite address locator you can perform 'one pass' geocoding using multiple address locators. This lets you use separate address locators for different geographic areas or different levels of detail to provide custom 'fall back' strategies to maximize the number of addresses that can be successfully geocoded.

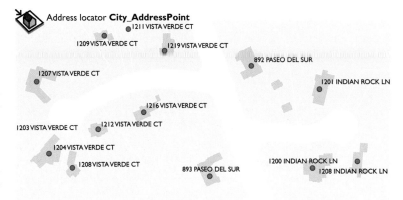

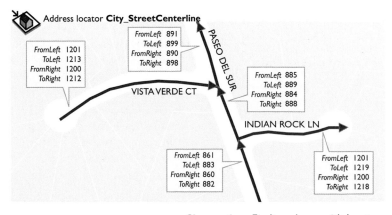

Address locators use a specific process to find a location. First, the address is parsed into individual address components. The address elements are then assigned to specific categories used in the search. The address locator searches the reference data to find potential candidates. Each potential candidate is assigned a score based on how closely it matches the address. Finally, the address is matched to the candidate with the best score.

In some geocoding applications, you will need to search for a location based on the intersection of two streets. When you add an address locator in ArcMap, you can specify a delimiter character such as '&' or '/' to denote a street intersection versus a street address. An example of an input street intersection would be "Hollywood Blvd & Vine St".

Sometimes streets have multiple names. A highway entering a city may be referenced as both 'Route 66' and 'Main Street'. To handle this possibility, you can define an alternate name table, so that if an address is not matched to a primary street name, the address locator can reference the alternate name table for other possible matches.

This chapter discusses the main concepts of address matching, but there are many more details to consider when building address locators, preparing reference data, and specifying geocoding rules for address matching.

For more information, go to http://support. esri.com and search on the term 'geocoding'. You will find white papers, a geocoding developer's guide, and a discussion forum for address management.

You will also find an ArcGIS Address Data Model that will be useful if your organization is responsible for developing the reference data for your address locator.

Once your address locator is built, you have several options for working with address data. You can find an individual address on the map, you can select a location on a map and find its address, and you can load an input table with many addresses and create a point feature class with the results.

Finding an individual address

You can use an address locator to find an individual address in ArcMap. To use an address locator in ArcMap, it must be loaded into the ArcMap document. You don't need to preserve the feature classes that you used to build the reference data for an address locator, but doing so will give you a visual reference to help you choose an appropriate candidate for an address. Remember that the reference data in an address locator is a snapshot in time, so to avoid confusion, use the same feature classes that you used to build the reference data in an address locator.

When you submit an address, you may get one or several candidates. You can visually inspect candidate addresses and select one to create a graphic point on the map.

Getting the address of a location

Sometimes you need to find the address of a certain location. In ArcMap, you can use the address inspector tool to point at a location and get the address of the location. This process is called reverse geocoding.

Geocoding a table of addresses

You can enter a table of addresses in any tabular format that ArcGIS can read. When you geocode a table of addresses, the address locator in your ArcMap document creates point features that represent the locations of the addresses.

Because address information is prone to error, a match score will be calculated for each address match. If the input address can be found without any ambiguity on the map, the match score will be 100. If some of your address elements have minor spelling errors (such as 'Valdes' versus 'Valdez') or an address element such as 'St' is missing, then the geocoder will probably find a match, but the match score will be about 70 or 80 percent.

When you geocode a table of addresses, you will specify a minimum match score. This is a threshold, perhaps 60 or 80, which will define the level of acceptable matches.

For addresses that cannot be matched, you can inspect each input address and either correct a misspelling, relax the spelling sensitivity, lower the minimum match score, or individually match the address to a point on the map.

The process of geocoding a table of addresses with an address locator is summarized on the facing page.

This illustration shows the three main steps in ArcMap for geocoding a table containing addresses.

First, you will add an address locator containing reference data. In this example, the address locator is based on address points containing attributes for address elements such as house number, street name and type, and direction prefix and suffix. The address points are typically located at the centroid of buildings, or if empty, the centroid of parcel lots.

1 Add an address locator

Add one or more address locators to an ArcMap document

Address locator **AddressPoint**

You begin geocoding by selecting an address locator and address table. You will specify an address field in the address table and set some parameters to control the matching process.

The matched addresses will become points in a point feature class which can either be static or dynamically linked to the address table. This feature class has fields containing the input and matched addresses, match scores, and several optional fields.

When you first geocode the address table, you will probably find that most addresses are successfully located on the map, some addresses are tentatively located and have ambiguous matches which need to be resolved, and a few addresses cannot be found.

2 Geocode addresses

Select an address locator
Select an address table and address field
Specify an output feature class for match points
Set geocoding options
- Spelling sensitivity
- Minimum candidate score
- Minimum match score
Set output options such as side and end offsets
Select output fields such as x, y coordinates and reference ID

Table **InputAddress**

AddressField
2006 Calle de Sebastian
2010 Conejo Dr
2001 Ft Union Drive
612 Calle do Leon
2115 Calle de Sebastian
504 Calle de Valdez
612 Calle del Valdez
2017 Valley Rio

504 CLL DE VALDES
2005 CLL DE SEBASTIAN
612 CLL DE LEON
2001 FORT UNION DR
2010 CONEJO DR
2115 CLL DE SEBASTIAN

After you geocode the address table, you can rematch addresses. Rematching addresses is an interactive process in which you can either relax geocoding options such as spelling sensitivity, correct misspellings in the input addresses, select address locations from a map, or choose from a list of match addresses similar to the input address.

Match scores are calculated for each matched address. Minor misspellings can be automatically corrected in the match address. Significant misspellings require inspection and correction. If a house number value cannot be found, you can choose from a list of addresses with a house number close to the input address.

3 Rematch addresses

Refine address matches by
- Adjusting spelling sensitivity
- Adjusting score values
- Correcting spelling errors
- Choosing tie matches on the map
- Picking address from the map
The geocoding result table shows both input and matched addresses with score values

Point feature class **GeocodingResult**

Score	AddressField	MatchAddress
64	2006 Calle de Sebastian	2005 CLL DE
100	2010 Conejo Dr	SEBASTIAN
100	2001 Ft Union Drive	2010 CONEJO DR
81	612 Calle do Leon	2001 FORT UNION DR
100	2115 Calle de Sebastian	612 CLL DE LEON
85	504 Calle de Valdez	2115 CLL DE
100	612 Calle del Valdez	SEBASTIAN
100	2017 Valley Rio	504 CLL DE VALDES

2017 VALLE RIO
504 CLL DE VALDES
2005 CLL DE SEBASTIAN
612 CLL DE LEON
2001 FORT UNION DR
2010 CONEJO DR
2115 CLL DE SEBASTIAN
612 CLL DE VALDES

How can insurance companies determine the number of people who might be impacted by living in a river's floodplain? What's the most efficient method to notify shoppers in an area about a new store opening? How do local residents find out about a new government service? How can address lists be maximized for effectiveness? Geocoding can provide answers to all these questions.

Geocoding is used to find the location of customers or assets and then understand the spatial relationships between the location and other geographic data, such as flood zones, school districts, or electoral boundaries.

Once you've geocoded one or many addresses, you can put those results to use for many applications. These are a few applications of geocoding at a typical city government.

Building permits

A common activity at city hall is applying for building permits. A builder or landowner comes to the permit counter and requests a building permit. The building permit clerk takes the address and using a desktop geocoding application, finds it on a map display. The clerk then examines that location on a map display and determines whether the permit request is compatible with the requirements of the districts containing the permit location. For example, buildings within historical districts must follow architectural guidelines and proposed construction within a flood plain must adhere to strict land-use restrictions. If the permit request passes zoning requirements, then information about local tax jurisdictions is entered and the permit is forwarded to an appropriate department for final approval.

Crime analysis

A city police department collects statistical data for all crimes, including crime type, date and time, and location by a street address or street intersection. Detectives use geocoding applications to plot the locations and times of crimes and search for patterns such as a series of burglaries. This information is used for heightened police patrols and investigations leading to the capture of criminal offenders.

This map shows one application of address geocoding. An issue before the Santa Fe city council was whether the downtown city hall office adequately served the administration of permits for short-term rentals of homes to visitors of the city. Tourism is an important part of the city's economy.

The GIS department at the city produced this map by geocoding a list of addresses from an administrative database table for short-term rental permits. The city council studied the distribution of short-term properties on this map and decided that the current office at the downtown city hall office was sufficient to serve the public administration of permits.

Address geocoding is a core tool used daily by local governments for efficiently delivering city services and for supporting decisions made by executive bodies at the city, county, and state level.

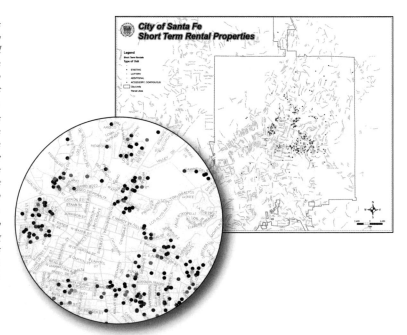

Public works department

The public works department at a city manages roads and several city utility networks. Whenever an activity is necessary on a city's infrastructure, it needs to be coordinated. For example, when a road cut is necessary for maintenance or repair of a water line, engineers locate that position using a geocoding application, determine whether other utility lines exist in close proximity, and synchronize that repair with any other maintenance needs. If there is to be a disruption in a public service such as water, the public works department will, using a geocoding application, identify the affected addresses and inform the residents by phone, mail, or signs.

Voting precincts

A city is subdivided into voting precincts for the election of city council members. As election day approaches, a frequent query to city hall is to ask for the voting precinct for a resident's address. When a call comes in, election staff enters the voter's address into a geocoding application, locates the precinct for the address, and informs the voter of the location for voting.

Enforcement of local codes

Some city codes have a geographic component based on distances. For example, a city code might specify that a new liquor store cannot be located within a 200-foot buffer distance around a school or religious building. When an application for such a license is made, a geocoding application can identify this type of spatial conflict.

Annexations and zoning changes

Whenever annexation results in changes to a city's boundary or land use zones are modified, the affected neighborhood must be notified. With a geocoding application, city staff can identify the properties affected by these changes, identify the owners and their addresses, and generate a mailing list to fulfill notification requirements.

Verification of regulatory compliance

Certain facilities within a city are subject to regulatory compliance. For example, cell phone antennas are often required to blend in with the local environment with minimal visual impact. A telecommunications operator must provide the city with addresses for these cell phone antenna sites. City inspectors will geocode these addresses on a map and then visit these locations to ensure conformity with local ordinances.

Geocoding applications are widespread

These are a few of the applications of geocoding at city governments. All types of organizations use geocoding. A political organization uses geocoding to locate volunteers and political donors. Retail businesses use geocoding to notify targeted customers of new store openings. Insurance companies uses geocoding to analyze risks in an area. Banks use geocoding to market financial services. Geocoding applications are widespread because addresses are the most common way that locations are specified. Geocoding lets you find locations from addresses and determine addresses for locations.

This map detail shows voting precincts with address points and voting locations.

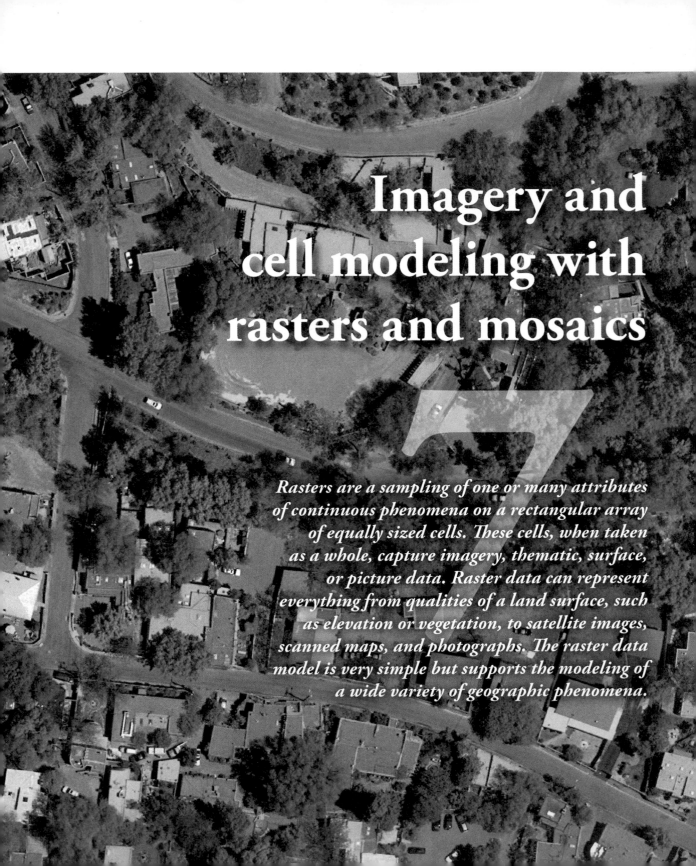

Imagery and cell modeling with rasters and mosaics

Rasters are a sampling of one or many attributes of continuous phenomena on a rectangular array of equally sized cells. These cells, when taken as a whole, capture imagery, thematic, surface, or picture data. Raster data can represent everything from qualities of a land surface, such as elevation or vegetation, to satellite images, scanned maps, and photographs. The raster data model is very simple but supports the modeling of a wide variety of geographic phenomena.

Every GIS user uses imagery directly and indirectly. Imagery is a core component of a geographic information system and enhances all of the other types of data. Some of the special capabilities that imagery brings to a GIS are:

- Imagery can provide a natural view of the world. Imagery, such as aerial photography or some satellite imagery, can provide an easily recognizable view of the earth's surface, therefore making an excellent informational layer to all users.

- Imagery is a great communicator. Since imagery is a picture of our world, it makes a great basemap layer for any type of map. People trust imagery because it is an actual view of the earth.

- Imagery presents new information. Much of the information about our environment—such as vegetative and soil type, distribution of pollutants, extent of urban development—is best detected using imagery.

- Imagery can be up-to-date. A key advantage of imagery is it can be the most current source of geographic data in a GIS. Many satellites and aerial sensors are constantly collecting imagery over parts of the earth every day. Some types of imagery, such as weather, are available in real time. This makes imagery especially valuable in times of disasters, such as flooding or fires. But this also makes it invaluable because it can be used to provide new and updated information, such as the discovery of a new road or loss of a wetland.

- Imagery can provide three-dimensional information. For decades, stereo images have provided topographic information and height information for building and other features. Advances in remote sensing and technology have led to using other forms of data, such as radar and lidar, to provide this same information on a global scale or local scale. When using specific high-resolution imagery or aerial photography, even to the naked eye, any user can clearly see the landforms on the earth such as mountains and canyons or the shape of three-dimensional features such as buildings and dams.

- Imagery is temporal. Imagery is a snapshot in time and can provide current and historical information.

Using imagery in ArcGIS

Imagery is used in two important and complementary ways in ArcGIS: as a background within a map and for image analysis. Imagery as a background adds a kind of "ground truth" that all map readers use as a reference for other geographic data displayed on a map. Image analysis involves processing the pixels in the imagery. It is done to obtain information such as the physical or biological characteristics of the earth, and the impact of human activities on the environment.

ArcGIS performs basic image analysis with focused imagery tools and fast image display capabilities. This allows intuitive and high-performance capabilities for navigating imagery integrated with map displays inside ArcMap.

ArcGIS lets you make the most of your investment in imagery with these capabilities:

- ArcGIS can use image data stored in many formats. Imagery can be stored within the geodatabase or accessed from all major image formats.

- ArcGIS has tools to improve and modify the appearance of the imagery, such as various standard or custom histogram stretches, setting transparency, and merging bands or changing band combinations.

- ArcGIS can process image data. Interactive tools and functions are available on the image analysis window, such as sharpening with a convolution filter or performing a difference calculation. More advanced data analysis tools are provided with the Spatial Analyst extension allowing you to derive new information from existing data.

- ArcGIS works with the leading image processing and analysis solutions for specialized applications.

- ArcGIS provides both desktop and Web solutions to make imagery available to users and provides the interface for those users to work with the imagery.

- ArcGIS can efficiently process and quickly serve huge amounts of imagery. ArcGIS manages large catalogs of imagery with a high-performance and scale architecture.

- ArcGIS can quickly incorporate new imagery. ArcGIS performs on-the-fly and dynamic server-based image processing that incorporates the most recently acquired image data.

- ArcGIS preserves information accuracy by dynamically mosaicking imagery while preserving the source image data and providing access to image metadata.

- ArcGIS provides support for enterprise interoperability and adheres to major information technology and geospatial standards.

Managing imagery in ArcGIS

How the image data is managed is as critical to a GIS as using the data. ArcGIS not only reads image data from various file formats, but provides the tools and data models to create, manage, and update vast amounts of image data, and provide this data to various applications, including ArcGIS, the Web, and more.

ArcGIS has the tools and technology to work with more than image data, but can also manage other data such as elevation data in point, TIN, and raster formats. Imagery is stored and accessed in ArcGIS in raster formats. Most raster data in a GIS represents imagery, but raster data also represents other types of data, such as terrain elevation and characteristics that are computed by geoprocessing operations, such as scoring land suitability.

It is recommended that you manage your image and raster in two parts:

- As raster datasets, which are a simple and flexible way to store single images either in the geodatabase or as files on disk. Each raster dataset generally represents a single dataset such as a satellite scene or set of images that have been combined into one.

- Using mosaic datasets, you can manage small to vast amounts of image and raster data. Mosaic datasets represent collections of images that can be easily accessed as a whole or as separate, individual rasters.

ArcGIS offers three ways to store raster data: raster datasets are a simple and flexible way to store single images, raster catalogs store multiple raster datasets in an easy to manage format, and mosaic datasets store multiple rasters along with predefined methods for merging and displaying imagery.

Use mosaic datasets if you are combining large amounts of imagery that are being continuously updated.

A single mosaic dataset was used to manage the image data used to create the image above. Using different functions, these three different representations of the image data have been created from a single source. From left to right: a sharpened near-infrared band combination, a normalized difference vegetation index created from the near-infrared and red bands, and a natural color pan-sharpened image.

Rasters sample a measured or calculated value on a regular matrix of cells.

Raster datasets represent geographic features by dividing the world into discrete square or rectangular cells laid out in a grid. Each cell has a location relative to an origin at the upper left corner of the raster grid and a cell value describing some entity being observed.

While the structure of raster data is very simple, it is exceptionally useful for a wide range of applications. Within a GIS, the uses of raster data fall in four main categories: rasters as basemaps to provide a photographic background display, rasters as surface maps to enable spatial analysis and relief mapping, rasters as thematic maps to present analytic results through derived classifications, and rasters as feature attributes to link a place with a photograph, document, or drawing.

The four main uses of raster data are summarized below with typical raster settings. These raster uses and their properties, display methods, and applications are discussed in more detail on the following pages.

Rasters as imagery

The most common use of raster data in a GIS is as a background display for other map layers. Images drawn below other layers provide the map user with confidence that map layers are spatially aligned and represent real objects. Three main sources for raster imagery are satellite imagery, orthophotos from aerial photography, and scanned maps.

Rasters as surface maps

Rasters are well suited for capturing any type of continuous value that forms a surface. Elevation values describing terrain are the most common application of surface maps, but other values such as rainfall, temperature, concentration, and population density can define a surface which can be spatially analyzed.

Rasters as thematic maps

A raster can comprise thematic data derived from analyzing other data. A common analysis application is classifying a satellite image by land-cover categories. This activity groups the values of multispectral data into classes (such as vegetation type) and assigns a categorical value. Thematic maps can also result from a geoprocessing model combining data from various sources, such as vector, raster, and terrain data.

Rasters as attributes

A raster can be a photograph, document, or drawing related to a geographic object or location. Rasters used as attributes can be stored as binary large objects (blobs) in a feature class table.

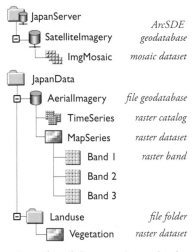

Raster data of all types can be stored in four basic ways: as a mosaic dataset, a raster dataset, a raster catalog, or a raster column in a feature class. Raster data can be directly stored in any type of geodatabase (file, personal, or ArcSDE) or accessed from raster files in the file system.

First, this chapter discusses properties of raster data common to all storage types. Later in this chapter, details and strategies for applying these raster storage options are covered.

Data source lists common sources for this type of raster.

Data type indicates whether cell values are continuously varying, as with imagery and surfaces, or categorical, as with thematic data.

Cell attributes summarize how cell values are stored: pixel depth, signed or unsigned, integer or floating, bands, value attribute tables, and colormaps.

Data structure means whether rasters are mosaicked together, managed in a catalog, or stored as a binary object in a table.

Display methods indicate how the rasters are typically displayed.

Compression settings indicate whether compression is lossless (for data on which analysis is done) and lossy (for display data).

Pyramid resampling recommends whether and how adjacent cell values should be averaged when resampled at a smaller scale. These are called pyramids, which are resampled rasters stored to aid display performance.

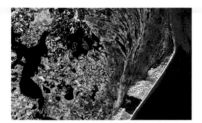

Data source	Satellite imagery, aerial photography, scanned maps
Data type	Continuous values
Cell attribute	One-to-many bands of 8-bit (or more) unsigned integers
Data structure	Mosaic dataset, single raster, or raster catalog
Display methods	RGB true-color or false-color composite, grayscale
Compression	Lossy or lossless
Pyramid resampling	Recommend cubic convolution or bilinear interpolation

Data source	Radar, conversion from terrain models, or lidar
Data type	Continuous values
Cell attribute	Floating values, complex
Data structure	Mosaic dataset, single raster, or raster catalog
Display methods	Hillshading, stretch colors on elevation, slope, shaded relief
Compression	Lossless
Pyramid resampling	Recommend cubic convolution or bilinear interpolation

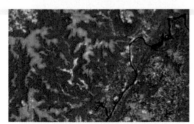

Data source	Land type processed from multispectral bands or converted
Data type	Categorical values
Cell attribute	8- or 16-bit unsigned integer values with raster attribute tables
Data structure	One raster dataset with a colormap
Display methods	Display RGB values from colormap
Compression	Lossless or none
Pyramid resampling	Recommend nearest neighbor

Data source	Digital cameras and document scanners
Data type	Continuous values
Cell attribute	Three bands of 8-bit unsigned integers or one 8-bit band
Data structure	Binary objects in feature class table
Display methods	RGB true-color composite or grayscale
Compression	Lossy
Pyramid resampling	Pyramids are usually not necessary

SATELLITE IMAGERY AND ORTHOPHOTOS

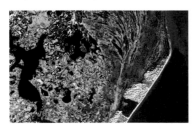

Advanced Very High Resolution Radiometer (AVHRR) sensors capture most of the world daily at a resolution of 1100 meters.

Landsat 7 image of Cabo San Antonio, Argentina, 15 meter resolution.

Half-meter resolution image of Khalifa Sports City, Doha, Qatar collected by GeoEye®-1 satellite.

Aerial orthophoto image of a hotel in Santa Fe, New Mexico. Resolution is six inches (15 centimeters).

Satellite imagery and orthophotos provide a photographic basemap on which all other layers are displayed.

Satellite imagery is continually collected along the earth's surface at resolutions from 1 km to less than 1 meter. Image data used for basemaps have cells on the order of 10 meters to 1 meter or less and have several or many bands of which three are displayed as an RGB composite or one as a grayscale.

Orthophotos are aerial photographs that have geometric rectification applied for camera tilt and surface relief. The result is a photograph map on which all true distances can be measured. These maps are typically imaged with a cell size from a few centimeters to one or two meters.

Data sources

Satellites image the earth from a wide range of orbital elevations with sensors capturing electromagnetic radiation at many wavelengths.

When you select satellite imagery for a basemap, three major considerations are resolution, spectral bands, and temporal coverage.

- What are the finest objects you need to resolve? Vehicles, streets, crop fields, and ecological areas require cell resolutions ranging from submeter to tens of meters.

- What spectral coverage is required? Are red, green, and blue bands for true-color display sufficient? Are other wavelengths such as infrared, radar, or thermal required for analysis?

- At which times and intervals do you need imagery? Do you need the most recent imagery? Do you need images by season or some specified time interval? Do you only want cloud-free imagery?

Digital aerial orthophotos are acquired by digital cameras or scanning aerial photographs and correcting for range differences caused by varying elevations.

An orthophoto series in widespread use is the digital orthophoto quadrangles (DOQ) from the U.S. Geological Survey, comprising rectified and mosaicked aerial photography, tiled at two sizes, 3.75 minutes square and 7.5 minutes square.

Display methods

Satellite imagery and orthophotos are commonly displayed three ways:

- With three bands representing red, green, and blue ground reflectances, forming true-color image maps.

- With three bands including one or more color reflectances not visible to the eye, such as infrared. These form false-color image maps.

- With three bands in which the red, green, and blue bands have been fused with a higher resolution band, to produce a pan-sharpened image map.

- With one band, to produce a grayscale image map.

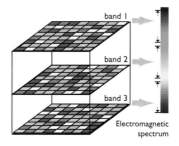

Displaying rasters in true-color uses blue reflectance for band 1, green reflectance for band 2, and red reflectance for band 3 for a combined RGB composite display.

Applications and analysis

Satellite imagery is commonly used for map background display. This is a cost effective way to add a base-map covering a large area.

Satellite images can be used in a photogrammetric system to derive features such as streets. Other raster products are derived from satellite imagery, such as inferring soil and vegetative units from multispectral reflectances.

Detail from a DOQ (digital orthophoto quadrangle), 1 meter resolution.

Orthophotos are an efficient base on which to compile a planimetric map with vector features for street edges, buildings, and facilities. Orthophotos are also analyzed to study land characteristics such as crop health.

Satellite imagery and orthophotos have applications in many domains such as agriculture, emergency response, geology, forestry, planning, transportation, and utilities.

Temporal modeling

Many satellites image locations on earth with intervals from one day to several weeks.

Landsat 7 images most places on earth about once every 16 days. Images like this can be viewed to observe changes on time-scales from days to years and analyze trends such as urban growth, deforestation, contamination, and ecological recovery.

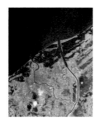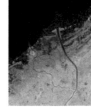

The city of Banda Aceh, Sumatra, before and after the tsunami of 2004. Landsat 7 images from May 14, 2004 and December 29, 2004.

Comparing orthophotos from different years reveals changes in the natural and built environments.

Data type and cell attributes

Most satellite and aerial imagery captures light reflectance from the ground in each sensor's spectral window. Since pixel values range from dark to bright, they are continuous values.

Satellite imagery and orthophotos used for background display three bands for true-color or false-color photography or one band for grayscale imagery.

Raster dataset structure

Satellite and aerial imagery used for basemap display are added to a mosaic dataset or mosaicked into a seamless raster dataset. Additionally, if images from various dates are used to monitor environmental changes, a mosaic dataset can be used to manage the separate raster datasets.

Compression and pyramids

Since satellite and aerial imagery are used as a map background, lossy compression can be employed to achieve high compaction. The compression level depends on the intended map-use scale and cell resolution. This value needs to be determined by careful visual inspection.

SCANNED MAPS

Sometimes, the best basemap is a published map that is scanned. Maps such as the USGS quadrangle series make popular basemaps because they offer a familiar and well-understood map background. A spatial reference can be attached to scanned maps to georeference them.

Data sources

Many government map series such as the U.S. Geological Survey quadrangle maps are available in scanned format. Some data providers scan these maps, trim them to neatlines, georeference the map sheet corners, and mosaic them across an administrative area such as state or province.

Maps are typically scanned at a resolution of 300 or 600 dpi (dots per inch). At this scan resolution, if the original map scale were 1:24,000, a scanned cell would have a resolution of about 2 or 1 meters, respectively.

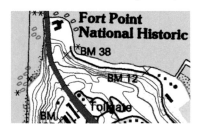

DRG (Digital Raster Graphics) images are scanned from USGS quadrangle maps, which show terrain, names, and major features.

Display methods

Color scan maps usually contain three bands for red, green, and blue reflectances and displayed as RGB composites, but cell values in a single band can also reference a colormap, a mapping between index values and RGB values to simulate the original appearance of the map.

Simple line maps that are displayed with one color may be scanned with a low pixel depth, such as 1-bit monochrome.

Scanned maps should be drawn at or near their published scale. This is because they contain text, line symbols, and area fills that can only display well at or near the original map scale.

Application and analysis

Scanned maps are an efficient way to utilize existing maps and charts. Using layer ordering and transparency settings, you can compare other map layers to scanned maps and validate data accuracy, currency, and completeness.

Temporal modeling

Organizations such as local governments and utilities have historic maps that can be scanned and added as a map layer. These maps provide valuable information about the construction history of facilities and patterns of landownership.

Scan of a historic parcel map can be overlaid with a modern map to analyze how parcel lots have been split, merged, or changed.

Data type and cell attributes

Scanned maps usually use 8-bit unsigned integer values to represent the color values displayed on the map. Cell values in a scanned map are continuous data values.

Scanned maps of simple line maps may be 1-bit monochrome, with each cell either black or white.

Scanned color maps can have one or three bands, typically stored as 8-bit unsigned integers, values from 0–255. If they have three bands, red, green, and blue values are stored for RGB composite display. If they have one band, they will have an associated colormap to assign RGB values to an index value.

Raster dataset structure

Scanned maps can be provided as individual mapsheets contained in a collection (like a mosaic dataset or raster catalog) or mosaicked into a single raster dataset. This depends on the spatial quality of the original maps and whether features at map edges match cleanly.

Compression and pyramids

A characteristic of scanned maps is that cell values are commonly identical from one cell to its neighbor. Many scanned maps have white or colored areas punctuated by lines and text. This allows a lossy compression.

Cubic convolution resampling

Bilinear interpolation resampling

Detail of a scanned map with pyramids generated with each resampling method

Nearest neighbor resampling

For scanned maps, bilinear interpolation and cubic convolution resampling methods should be compared under close visual inspection.

TERRAIN MAPS

Rasters model the surface of the earth with cell values for elevation. These rasters are known as digital elevation models (DEM), digital terrain models (DTM), and digital surface models (DSM).

Many analytic tools are available for computing watersheds and drainage networks, viewsheds, slope, and insolation. Display methods combine hillshading, elevation, and slope to display realistic landscapes.

Data sources

Elevation models can be generated by stereo orthophotos, interferometric radar, lidar, sonar, or even GPS coordinates. Raw elevation data is usually not in a grid form, but elevations are resampled on raster cells.

The USGS National Elevation Dataset (NED) provides seamless elevation for the U.S. at two arcsecond resolution.

Digital Elevation Models (DEM) are a USGS dataset at several scales that samples elevations on a raster. Elevation data are available for most of the world at about 100-meter resolution from NASA's Shuttle Radar Topography Mission (SRTM).

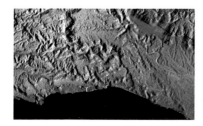

Coastline of Oman from SRTM data

Lidar and sonar elevation models can generate very high resolution data, often less than 1 meter.

Lidar data from a highway interchange

Lidar data can build an extremely detailed surface model, including objects on the earth's surface such as buildings and vegetation.

Display methods

For elevation data, a stretch renderer or color ramp is commonly used to portray hypsometric colors for different elevation ranges. Hillshaded relief maps can be generated from elevation data, combined with slope, aspect, and elevation.

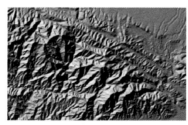

Hillshading on a raster with elevations

For each cell in a digital elevation model, a hillshade map draws shades that simulate the illumination of a surface, based on a specified sun direction and angle between the sun and the slope of the local surface. Hillshade maps work by providing visual cues for depth perception. Hillshade maps are drawn with sun illumination from a specified sky direction.

This map shows the slope of a terrain. The red cells show steep areas and the green cells show flat areas.

Slope is an important consideration for selecting a building site. Many localities have ordinances specifying allowable slope values for construction.

This map shows elevation by shaded colors. The red cells show lower elevations and blue cells show higher elevations.

While shaded elevation maps look flat by comparison to hillshade maps, they do highlight zones of common elevations.

This map shows shaded colors for elevation combined with hillshading.

The combination of these display methods creates an attractive map that simultaneously shows heights and the surface shape.

Applications and analysis

Elevation rasters have many uses.

- Hydrologists apply analytic methods on DEMs to compute watersheds and stream network.
- Military planners plan flight paths for reconnaissance from surface models.
- Local governments use slope analysis to enforce building codes on hillsides.
- Civil engineers perform volumetric calculations to balance cut and fill on road projects.
- Communication utilities use line-of-sight analysis for locating cell phone towers.
- Companies and agencies that operate aerial photography and satellite imagery use DEMs to correct for scale distortion in their map product.
- Cartographers use hillshading and hypsometric colors to add depth to a map.

An important and interesting application of raster surfaces is hydrology. The hydrologist is interested in estimating how a severe storm event will impact a stream network.

Elevations displayed from a DEM

The analysis starts with a raster surface from DEM data—raster geoprocessing tools detect ridges and stream lines to derive vector features for a stream network.

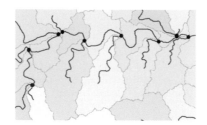

Catchments and stream reaches are derived from a DEM.

Once the stream network is built, rainfall amounts can be added from precipitation maps recorded by weather radar systems. A hydrologist analyzes how rainfall accumulates on a stream network to predict flood levels, time of surge, and areas of inundation.

Temporal modeling

One area of temporal analysis with elevation rasters is the monitoring of volcanically active areas.

Subtle elevation shifts on the order of centimeters can be measured by comparing radar returns made at different dates. These measurements are important for the safety of local residents; bulges in the land may be an indicator of underground magma accumulation.

Data type and cell attributes

Cell values in a raster of elevation have a zero point at a geodetic sea level with continuously incrementing values, or decrementing for bathymetric surfaces. Cell values in an elevation raster are continuous.

Elevation cell values usually are floating point values. Elevation values are usually the same as the map unit: meters, feet, or other.

Compression and pyramids

Lossless compression such as LZ77 is required because elevations are used in analysis and must retain their precise values.

Cubic convolution resampling of pyramids is recommended for a smooth, crisp appearance.

This map displays temporal variations in sea surface temperature between January and July. These measurements are made by the NASA Earth Observations program.

CONTINUOUS SURFACE MAPS

Elevations are just one of the surfaces that can be modeled with rasters. Other continuous surfaces modeled with rasters include rainfall, temperature, contaminant level, groundwater level, noise, gravity, and population density. Continuous surfaces can be captured through sensors such as radiometers, derived from geostatistical analysis, or calculated from a geoprocessing model.

Display methods

Continuous surfaces have a single cell value and can be displayed with a stretch, classified, grayscale, or with hillshading display method.

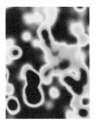

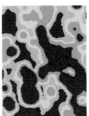

Population density data drawn with the stretch display method and the classified display method

The stretch display method applies a color ramp to the range of elevation values. The classified display method applies a color for all cell values within a classification.

Applications and analysis

Continuous surfaces model a wide range of field phenomena.

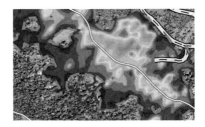

Bauxite ore concentration in Jamaica

Oil contaminant plume in an aquifer, modeled using groundwater level, soil porosity, contaminant surface tension and viscosity, and other factors

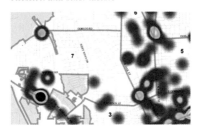

Crime density map highlighting burglary hot spots

Temporal modeling

Many continuous surfaces change over time. Contaminant plumes spread from a source, precipitation falls with the passing of a storm, and temperature is a field continuously varying in space and time.

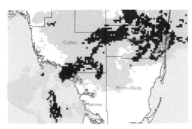

Nexrad radar rainfall data for southern Florida, 15 cumulative minutes.

Rasters with time-variant surface values can be stored as separate raster datasets for each point in time. These raster datasets can be managed using a raster catalog.

Data sources

Continuous surfaces have several types of data sources:

- A surface can be remotely sensed, such as a map of ocean temperatures collected from satellites.
- A surface can be derived from geostatistical analysis on a set of input points, such as population density built from city centroid points and population counts.
- A surface can be calculated from a geoprocessing model, such as estimating jet noise levels from factors such as flight paths, terrain, and atmospheric condition.

Government weather agencies and commercial weather data services provide near real-time precipitation maps from radar or satellite sensor.

Data type and cell attributes

Cell values in a surface raster are continuous data values. Surface cell values usually are floating point values. The measurement unit can be degrees on the Celsius scale, cm of rainfall, decibels, parts per million of contaminant, or other.

Raster dataset structure

Before analysis, surfaces representing a point in time or permanent state would generally be mosaicked together. If surface values are time-variant, they are stored as separate raster datasets in a raster catalog.

Compression and pyramids

Additional analysis would usually be performed on this data, and thus, original values must be preserved. Using a lossless compression (LZ77) would be most appropriate. Pyramids for surfaces should use nearest neighbor resampling to preserve the original data.

LAND CLASSIFICATION MAPS

Values for raster cells can represent a category or classification of data, such as landownership type or vegetative type. The value from one cell to another is either identical or changes abruptly. This type of data appears as a set of zonal regions with common values, such as land-use maps or forest stands.

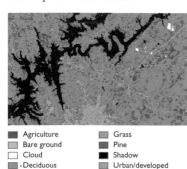

Agriculture
Bare ground
Cloud
Deciduous
Deciduous/pine mixed
Grass
Pine
Shadow
Urban/developed
Water

Cell values for spatially discrete data represent a classification that applies to the full area of a cell.

Display methods

Thematic maps have well-defined colors that may or may not resemble the natural feature color but provide a key for the classification of land units.

		red	green	blue
1	↔	255	255	0
2	↔	64	0	128
3	↔	255	32	32
4	↔	128	255	128
5	↔	0	0	255

Thematic maps can be displayed with colormaps, which is an assignment of an RGB value to a list value.

Applications and analysis

Some applications of land classification maps include predicting development impact, estimating wildlife habitat areas, developing remediation plans, inventorying forest stands, prospecting for mineral ores, and marketing analysis.

Temporal modeling

Land classification rasters made over time can be compared to reveal rates of urban encroachment, shifts in ecological zones, and forest harvesting.

Maps showing past, present, and projected land status can be influential in public policy debates on land management.

Data sources

Land classification maps are usually derived by matching spectral signatures against known reflectance patterns for vegetative or soil types.

The multispectral sensors on satellites, such as Landsat 7 and EnviSat, permit sophisticated identification and classification of land cover units, based on matching measured spectral responses at several or many spectral bands.

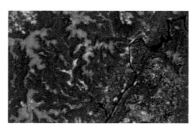

Multiband imagery from Landsat and similar environmental satellites are the primary input for processing land classification maps.

Land classification maps can also be made by converting vector data, such as soil type polygons.

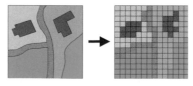

A thematic raster map can be converted from a polygon feature class.

Thematic rasters can be derived from other data sources. They may be the result of a geoprocessing model combining many data sources (such as elevation, access to water, and population density), a conversion from vector or terrain data, spectral classification, or the result of a raster operation performed on one or many raster sources.

Data type and cell attributes

Thematic and categorical maps have cell values that are derived, arbitrary, have no numeric meaning, and represent an index value in a list. Thematic cells have discrete data values.

Thematic maps are typically 8-bit data because 0–255 is sufficient for most categorical data. Colormaps are often used for thematic maps because they are an efficient way to map RGB colors to thematic categories.

Raster dataset structure

When the land classification values are consistent, you can mosaic the datasets together. If they are not consistent, the raster datasets need to be kept separate and are best stored in a raster catalog.

Compression and pyramids

Additional analysis would usually be performed on this data, and thus, original values must be preserved. Using no compression or a lossless compression such as LZ77 would be most appropriate.

Nearest neighbor is used for pyramid resampling because it maintains cell values. Averaging numbers which represent categorical values does not make sense.

PHOTOGRAPHS, DRAWINGS, AND DOCUMENTS

Besides images of land, rasters are also used for photographs of features, drawings about features, and documents describing features. Rasters as attributes are an important augmentation to the information presented in a map.

Rasters as photographs of features can be captured by digital cameras, which make pictures in JPEG format with red green blue colors.

Electric utility transformer photo, linked to a transformer point feature

Photographs can be an attribute of point features, thus describing locations, but photographs do not have a spatial reference. A video clip can also be an attribute of a feature.

Features on a map can also have associated drawings and documents as attributes.

Drawings may be building blueprints, site maps, or a wiring diagram for a switch box.

These types of drawings are not spatially referenced, but they are georeferenced—tied to a place on Earth. Features such as land parcels have associated legal documents that contain the precise description of land.

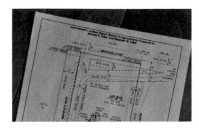

Documents may be property deed records, work orders, and maintenance records.

Legal documents can be scanned and stored as an attribute value of the parcel polygon feature.

Display methods

Digital cameras record red, green, and blue values, so RGB Composite display is used for color photographs. Drawings and documents may be scanned in monochrome, 8-bit grayscale, or color. Grayscale or RGB composite display is used.

Applications and analysis

Rasters as attributes are used anytime a user wants to attach a picture to features in a feature class.

When biologists develop an environmental impact assessment, photographs can document habitat conditions at a number of locations.

A city government can photograph municipal buildings and other facilities and georeference them to building polygon features.

Hydrologists studying watersheds can attach pictures of erosion, sloping hillside, or mouth of a river.

Temporal modeling

Photographs and documents record a condition at a point in time. If that point in time is recorded or can be derived, then drawings and documents can be retrieved by a combined spatial and temporal query.

Landownership documents contain transactional details about how parcels are split, merged, or modified through time.

Photographs of a landscape taken through decades show whether vegetative species are advancing or receding.

Data sources

Photographs, drawings, and documents can be scanned or taken with a digital camera.

Data types and cell attributes

Photographs capture light reflectance in red, green, and blue. Since pixel values range from dark to bright, they are ratio data values.

Photographs have three bands for red, green, and blue. Most commonly, cell values in each band are stored as 8-bit unsigned integers, with a range of 0 to 255.

Raster dataset structure

Rasters that are attributes of a location or feature are stored in their native format (JPEG, TIFF, or other) as binary objects in a feature class table. Each feature in a feature class can have one raster as an attribute.

Compression and pyramids

Photographs use lossy compression, such as the JPEG format, which can produce high visual quality with substantial compression ratios. Pyramids are not usually created for photographs. Scanned drawings and documents can captured in monochrome, grayscale, and color. Pixel values range from dark to bright and are ratio data values.

Pyramids are not usually created for scanned drawings and documents.

All four broad uses of rasters are modeled with a raster dataset, stored inside a geodatabase or a raster file. The remainder of this chapter discusses the properties and structure of raster datasets.

A raster dataset is a geographic dataset with square or rectangular cells (called pixels) that tesselate a rectangular area.

Data source

A raster dataset has a name, type, and location. There are three types of rasters: file system, image service, and geodatabase. A file system raster is recognized in ArcCatalog if it is in one of a number of recognized file raster formats.

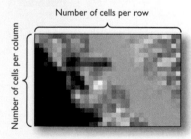

Raster datasets can be stored in a geodatabase or referenced.

Rows and columns

A raster dataset has cells organized into rows and columns.

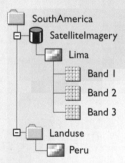

Number of cells per row

Number of cells per column

The number of cells is the product of the number of rows and the number of columns.

Raster bands

A raster dataset can have one or more raster dataset bands. Bands are used for sensors that capture an image at several wavelengths simultaneously.

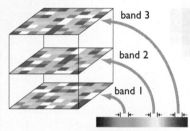

band 3
band 2
band 1

Electromagnetic spectrum

Three-band rasters usually represent red, green, and blue wavelengths which are combined to make natural-color images. Some imaging satellites such as Landsat 7 record spectral reflectance at 6 or more bands.

Cell properties

Each cell has a width and height, which is considered the resolution of the raster data.

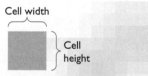

Cell width

Cell height

The uncompressed size of a raster dataset is determined by the number of cells, how many bits are required for cell values, and how many bands are present.

Format

Many raster formats in widespread use are recognized, such as DIGEST ASRP, DTED, IMG, JPEG, MrSID®, NITF, PNG, TIFF, and others.

Pixel depth and type

The number of bits used for sampling a value at each cell ranges from 1 to 64 bits. The number of values is two to the power of the number of bits. Cell values can also be signed and unsigned, integer, or floating point.

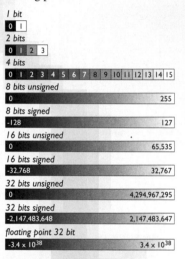

1 bit
0 1

2 bits
0 1 2 3

4 bits
0 1 2 3 4 5 6 7 8 9 10 11 12 13 14 15

8 bits unsigned
0   255

8 bits signed
-128   127

16 bits unsigned
0   65,535

16 bits signed
-32,768   32,767

32 bits unsigned
0   4,294,967,295

32 bits signed
-2,147,483,648   2,147,483,647

floating point 32 bit
-3.4×10^{38} 3.4×10^{38}

Many rasters use 8-bit unsigned integer cell values for single-band grayscale imagery, as well as three-band RGB rasters. 1-bit rasters are black and white images, used for scanned line drawings.

NoData value

While the cells in a raster dataset necessarily span all the rows and columns forming a rectangular area, not every cell will have a data value. Operations such as coordinate transformation and mosaicking will create cells with no data values. An arbitrary value, such as -9999, may represent no data as well as the NoData value.

Colormaps

Colormaps can be applied to single-band rasters with integer values of 8- or 16-bit depth. A colormap is a simple table that relates a set of integer cell values to defined RGB colors. Cell values represent a categorical value, such as land-use, vegetative types, and geological units

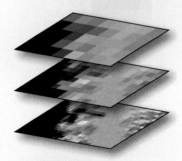

If a raster has a colormap, pyramid interpolation will only apply nearest-neighbor interpolation so that categorical values are not averaged.

Pyramids

Pyramids improve display performance of rasters at varying map scales.

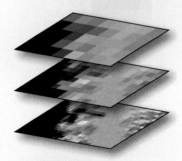

Reduced-resolution pyramids can be optionally created for a raster to optimize display performance at varying map scales. A pyramid is a set of progressively resampled rasters at each level. When the resampling factor is 2 (a 2-by-2 cell matrix) four cells are resampled to one cell. If using a factor of 3, a 3-by-3 cell

matrix is used and nine cells are resampled to one. Using a factor of 2 is optimal for speed; however, a factor of 3 often provides adequate display speed and can reduce the overall storage required for the pyramids because fewer are generated.

Compression

Raster dataset formats often support several data compression methods, including lossless, lossy, and wavelet.

Lossless compression reduces the file size of the raster dataset, but no cell information is lost. Use lossless compression if you are performing raster analysis.

Lossy compression reduces the file size to a greater degree, but with some loss of information content. Use lossy compression if you are using raster datasets for a display background and if you can tolerate slight image degradation.

Extent

A raster's extent is the top, bottom, left, and right coordinates of the rectangular area covered by a raster.

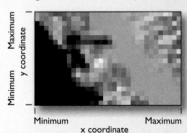

The area covered by a raster dataset is the product of the differences of minimum and maximum x and y coordinates.

Spatial reference

Like other geographic datasets, a raster dataset has a spatial reference, which is defined as a coordinate system, units of measure, and a precision value for internal coordinate storage.

Because raster datasets have spatial references, they can be overlaid with other geographic data, mosaicked with other raster datasets, and serve as a background to other map data.

Statistics

Statistics are properties calculated on each raster dataset band to enable layer drawing methods such as the raster stretch renderer.

The minimum, maximum, and mean values of cells and the standard deviation of distribution are calculated and stored as raster dataset properties.

For very large rasters, a small sampling of cells may be sufficient. You can specify how many cell rows and columns to skip for deriving statistics and this will drastically reduce the time to calculate statistics on large rasters. You can also specify a value to ignore in creating statistics.

Raster datasets are two-dimensional arrays of cells (or pixels). Each cell's height and width are the same throughout the raster dataset, and cells may be square or rectangular.

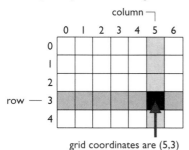

grid coordinates are (5,3)

Each cell in a raster is uniquely referenced by its column and row value. You can think of row and column values as integer coordinates of a grid, but unlike the Cartesian coordinate system, row values start at 0 at the top of the raster and increase in value in a downward direction. The rows and columns of raster data are always parallel to the x- and y-axes.

Cell sampling

Cell values can represent a value of a point or area.

For certain types of data, such as elevation surfaces, cell values represent a measured (or interpolated) value at the center of the cell.

For maps showing land classification attribute, the cell value may represent the dominant category.

Cell resolution

The cell size determines the resolution of raster data. Features identifiable in a raster cannot be located more precisely than the resolution of the raster.

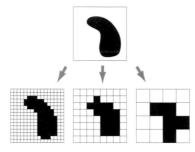

The size chosen for a grid cell of a study area depends upon the data resolution required for the most detailed analysis. The cell must be small enough to capture the required detail, but large enough so that computer storage and analysis can be performed efficiently.

Cell resolution also determines the largest map scale at which this data can be properly displayed.

Cell attributes

The value associated with a cell defines the class, group, category, or measure at the cell position. Cell values are numbers, either integer or floating point.

Cell values can represent a measured or derived quality of a location, such as reflectance, density, precipitation, or elevation. These values are either integer for reflectance or floating point for many other measurements.

Cell values can also represent categorical values such as land classification, which may be calculated from a geoprocessing model with many input geographic datasets. These values are always integer numbers.

Cells can also have a no data value to represent the absence of data.

While each cell has a single value and a color has three components (red, green, and blue values), color is associated with a raster most commonly by using raster bands for RGB composite display, or in the case of thematic maps, colormaps that associate an RGB color with a thematic index value.

Extending cell attributes

Rasters that have integer valued cells can be extended with an optional raster attribute table, which records attributes for each unique cell value. You can add custom fields to this value attribute table.

	Value	Count	Type	Code
	23	7	Fir	400
	29	18	Juniper	410
	31	10	Aspen	420
	37	18	Piñon	500
	41	4	Cottonwood	510
	43	7	Walnut	600

For rasters in which cell values represent a category, this value may be a code for a richer classification. For example, a map of tree types may have code '23' for fir, '29' for juniper, and so on. A raster attribute table can store additional information about this unique value, such as how many times it occurs in the raster, descriptions, and other codes.

Understand that with a raster attribute table, attributes are not being stored for individual cells but for unique cell values.

Types of data

Data can be considered to be one of four types: nominal, ordinal, interval, and ratio. Being clear about the type of cell data you are modeling will guide choices about cell attributes, compression, pyramid resampling, and map display.

Nominal data

A value of nominal data identifies one entity from another. These values establish the group, class, member, or category with which the geographic entity at the position of the cell is associated. These values are qualities, not quantities, with no relation to a fixed point or a linear scale. Coding schemes for land use, soil types, or any other attribute qualify as a nominal measurement.

Nominal data values are categorized and have names. The data value is an arbitrary type code.

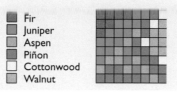

- Fir
- Juniper
- Aspen
- Piñon
- Cottonwood
- Walnut

integer cell values

lossless or no compression

nearest neighbor pyramid resampling

display by colormap

Ordinal data

A value of ordinal data determines the rank of an entity versus other entities. Examples are land suitability classifications and soil drainage rank. These measurements show place, such as first, second, or third, but they do not establish magnitude or relative proportions. You cannot infer a quantitative difference, such as how much an entity is larger, higher, or denser than the others.

Ordinal data values are categorized, have names, and the value is in a numerical rank.

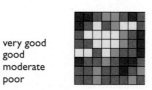

- very good
- good
- moderate
- poor

integer cell values

lossless or no compression

nearest neighbor pyramid resampling

display by colormap or classification and color ramp

Interval data

A value of interval data represents a measurement on a scale such as time of day, spectral range, temperature in Fahrenheit degrees, voltage potential, and pH value. These values are on a calibrated scale but are not relative to a true zero point. You can make relative comparisons between interval data, but their measure is not meaningful when compared to the zero point of the scale.

Interval data values are numerically ordered and the interval difference is meaningful.

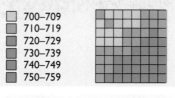

- 700–709
- 710–719
- 720–729
- 730–739
- 740–749
- 750–759

floating cell values

lossy compression for display, lossless for analysis

bilinear interpolation or cubic convolution pyramid resampling

display by grayscale, RGB composite, stretch, or classification

Ratio data

A value of ratio data represents a measure on a scale with a fixed and meaningful zero point. Mathematical operations can be used on these values with predictable and meaningful results. Examples of ratio measurements are age, rainfall, population, distance, weight, and volume.

Ratio data values measure a continuous phenomenon with a natural zero point.

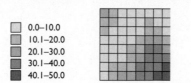

- 0.0–10.0
- 10.1–20.0
- 20.1–30.0
- 30.1–40.0
- 40.1–50.0

floating cell values

lossy compression for display, lossless for analysis

bilinear interpolation or cubic convolution pyramid resampling

display by grayscale, RGB composite, stretch, or classification

Choosing methods for displaying rasters depends on the data contained in the rasters and what you are trying to show.

Rasters with multiple bands can have three bands displayed as an RGB composite. Color satellite imagery and orthophotos are displayed this way.

Rasters that have single bands have a variety of display methods available. If cell values represent thematic groupings, a colormap can assign RGB colors to a thematic index value. If cell values represent continuous values, values can be assigned to a continuous ramp of colors for smooth display. Continuous cell values can also be colored by classifications of values.

Displaying color images

Color imagery is displayed using three raster bands forming an RGB (red, green, and blue) composite.

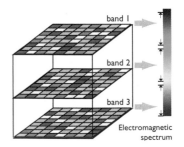

Orthophoto displayed as an RGB composite.

For true-color display, band 1 has blue reflectance, band 2 has green reflectance, and band 3 has red reflectance.

False-color images are useful for analyzing vegetation and geology. Instead of using red, green, and blue bands in an RGB composite, one or more bands, such as infrared, are substituted.

Same orthophoto but with an infrared band in the RGB composite.

Displaying single-band rasters

Rasters with single bands can be drawn in three ways: with colors applied continuously through the range of values, with a set of discrete colors for classifications of values, and with a random or assigned color for each unique value.

Single-band grayscale display of orthophoto.

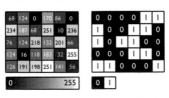

Grayscale images usually have cell values ranging from 0 to 255 and are used for black & white aerial photography. Monochrome images have two cell values, 0 and 1, and are used for simple scanned drawings.

Simple grayscale display of single-band rasters is done through the stretch display method, using a color ramp from white to black and no stretch specified. Monochrome display is done the same way, except only two values, 0 and 1, are used.

Stretching single-band rasters

A single-band raster with continuous values can be displayed by applying a color ramp to cell values. Several methods are available to stretch the assignment between colors in the color ramp and cell values. Stretch display on rasters works well for grayscale imagery and surfaces.

 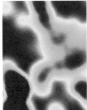

Stretch display with standard deviation.

 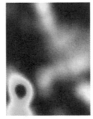

Stretch display with histogram equalize

 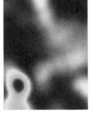

Stretch display with minimum maximum

Classifying single-band rasters

Rasters can be displayed by drawing range classifications of cell values with defined colors. While losing a smooth appearance, classified display of rasters is effective for applications with cut-off values, such as suitability analysis.

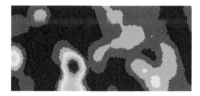

Classified display by defined interval—each classification is divided by regular values such as 1000, 2000, and so on.

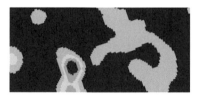

Classified raster display by equal intervals— the range of cell values is divided into equally sized classes.

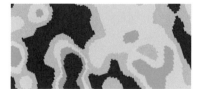

Classified raster display by natural breaks— natural groupings of cell values are found and classified.

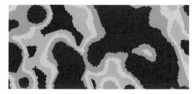

Classified raster display by quantile—each class has the same number of cells.

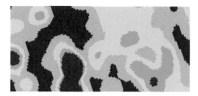

Classified raster display by standard deviation—each classification, based on cell values, vary from the mean.

Displaying thematic maps

Values in a raster dataset can be uniquely drawn. That is, each unique cell value can be assigned a randomly assigned color that you can modify. If your raster dataset has a colormap, it will be recognized and used for display.

Thematic land map with colormap.

								red	green	blue
1	5	3	2	2	4		1 ↔	255	255	0
5	2	4	2	5	1		2 ↔	64	0	128
5	5	5	5	3	3		3 ↔	255	32	32
2	1	2	4	1	3		4 ↔	128	255	128
4	4	4	1	1	3		5 ↔	0	0	255
2	4	2	1	3	3					

A colormap is simply a table referencing thematic index values to red, green, and blue values for map display.

Displaying surfaces

If cell values in a raster represent a surface elevation, you can create hillshaded relief. When a raster is hillshaded, a new raster is created with calculated shade values for the cells.

Hillshading is commonly combined with stretch rendering of a surface to display graduated colors for elevation on top of shading for landscape relief.

This terrain map combines hillshading with color display of elevation and slope.

Displaying mosaic datasets

When raster datasets are managed by a mosaic dataset, there are default and customizable rules to dynamically create mosaicked images, as well as applying on-the-fly processing, such as orthorectification, color correction, and image algebra. Additionally, you can view the mosaicked image and the footprints for each image.

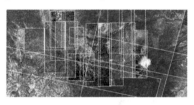

Mosaic datasets have embedded rules to intelligently fuse many rasters into a seamless display. The source rasters are not changed and you can update rasters any time.

Displaying raster catalogs

When raster datasets are managed by raster catalogs, you can define whether the raster tiles are displayed as images only, images with wireframe, or wireframe only. The wireframe display lets you see where the coverage of each image is in the raster catalog.

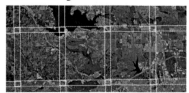

Raster catalog shown with wireframe outline.

With large rasters in a geodatabase, two factors in your control greatly affect display performance: the use of pyramids and compression. You have a range of choices with regard to each of these capabilities, and you may need to consider several factors and the application requirements to make the best use of these choices.

The primary benefit of compressing your data is to reduce storage space. An added benefit is greatly improved performance because you are transferring fewer packets of data from the server to the client application.

Data compression affects the tiles of raster data before storing them in the geodatabase. The compression can be lossy (JPEG and JPEG 2000), or lossless (LZ77). Lossless compression means that the values of cells in the raster dataset are not changed or lost. The amount of reduction in data size will depend on the type of pixel data; the more homogeneous the image, the higher the compression ratio.

Lossy compression is well suited for background images not intended for analysis. It results in faster data loading and retrieval. Less storage space is needed, since the compression ratios can be 5:1 or 10:1 with JPEG and up to 20:1 with JPEG 2000.

Lossless compression should be chosen for the following reasons:

- The raster datasets are to be used for deriving new data or for visual analysis.

- The required compression is no more than 3:1.

- You want to preserve the information content of the original data.

- Your inputs have already been lossy compressed.

JPEG compression by map scale

Percentages reflect the target amount of original data preserved by the algorithm.
Higher percentages are needed at larger map scales (smaller denominators).

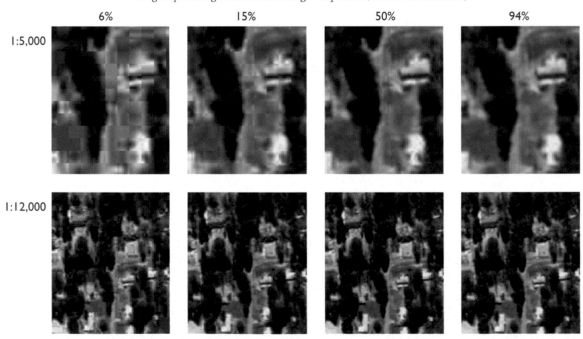

	6%	15%	50%	94%
1:5,000				
1:12,000				

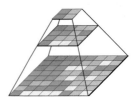

It is important to prototype to determine the most appropriate resampling technique for your data. Remember that pyramid resampling only affects the display, not the original data.

Pyramids are reduced-resolution representations of your dataset used to improve performance. Pyramids speed up the display of raster data by varying the level of detail in accordance with the extent of data to be viewed. When you view an entire dataset, the level of detail need not be very great. When you zoom to a small area, the detail must be as finely resolved as possible. The database server chooses the most appropriate pyramid level automatically, based on the user's display scale.

Pyramids are created by resampling the original data into a set of lower-resolution representations. When you create pyramids, additional storage space is required to store the resampled data. When you enable pyramiding for a raster dataset, a key parameter to consider is the resampling method to be used. Nearest neighbor should be used for nominal data or raster datasets with colormaps, such as land-use or pseudocolor images. Bilinear interpolation or cubic convolution should be used for continuous data, such as satellite imagery or aerial photography.

Nearest neighbor

The nearest neighbor resampling method determines the new cell value by transferring a value from the closest cell in the origin raster. This method has the virtue of being quickly calculated and preserving the range of attribute values. When the cell attribute represents a categorical value, such as land-use type, this is the method that should be used; otherwise, integer code values will be averaged, yielding invalid results. The disadvantage of this method is that a stair-step effect is sometimes noticeable, especially along linear features such as roads with high-contrast background.

Bilinear interpolation

The bilinear interpolation resampling method uses the four closest cells in the origin to calculate a geometrically weighted average for the resampled cell value in the resultant raster. The advantage of this method is that the stair-step effect, visible with the nearest neighbor method, is eliminated and images look smoother. The disadvantage of this method is that very small features with high contrast relative to neighboring features have their cell values averaged, so these small features sometimes become indistinct or ambiguous.

Cubic convolution

The cubic convolution resampling method uses the sixteen closest cells in the origin raster to calculate a geometrically weighted average attribute value for the resampled cell in the resultant raster. This method yields similar results to bilinear interpolation, but smooths the cell values to an even greater extent. Cubic convolution is suitable for rasters used as an image background in a GIS, but should not be used for rasters that represent a classification of values, such as land-use types.

When a raster is captured with devices like satellite imaging systems or desktop scanners, the raw data is just rows and columns of cells. In order to use such data in a GIS, either to draw on the screen with other data or overlay in an analysis operation, the data must be in a common coordinate system. This is a real-world coordinate system.

Georeferencing

Georeferencing is the process of establishing a relationship between the raster's (row, column) coordinate system, sometimes called image space, and a real-world (x,y) coordinate system, called map space. A similar transformation process occurs when establishing the relationship of feature data in one coordinate system (i.e., digitizer units) to another map coordinate system.

You can define a transformation to register the raster to real-world coordinates. This lets your raster in the same space as your other geographic data, such as vector features or a surface in a tin dataset.

To georeference a raster, it needs to be registered to a coverage, map, or set of coordinates that are in map space. The registration process is normally an interactive one where common locations are selected both in the raster and the other geographic dataset. For imagery, these are normally things like road intersections that are easily identifiable in both datasets. Once the common locations are selected, a polynomial transformation is built to model the scale, rotation, and skew between the two coordinate systems.

The georeferencing information is stored internally to some raster formats, or in external files such as the raster auxiliary file or the world file for other formats such as JPEG or BMP.

Using this information, the raster can be transformed on-the-fly and drawn in the map space of your other data. If you have also stored the map projection information, it can also be projected into other coordinate systems.

Rectification

To align an image axis with a map-space axis, an image must be rectified by resampling it based on the transformation built during the registration process. In resampling, a mesh is overlaid on the raster and a value is assigned to each cell according to the center's proximity to the values of the centers of the cells in the rotated raster. The values assigned to the output raster will be determined by the type of resampling: nearest neighbor assignment, bilinear interpolation, or cubic convolution.

You may wish to rectify a raster to remove skew or rotation, to orient its cells orthogonally to the map orientation. The primary reason not to rectify is because any such resampling of a raster will induce a small amount of error. This amount of error is not something you can see but can be important in multispectral analysis where minute differences in cell values can be significant.

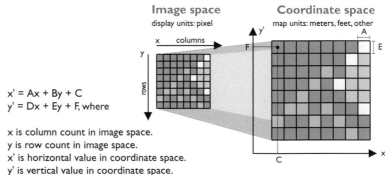

$$x' = Ax + By + C$$
$$y' = Dx + Ey + F, \text{ where}$$

x is column count in image space.
y is row count in image space.
x' is horizontal value in coordinate space.
y' is vertical value in coordinate space.

A is width of cell in map units.
B is a rotation term.
C is the x' value of the center of upper-right cell.
D is a rotation term.
E is negative of height of cell in map units.
F is the y' value of the center of upper-right cell.

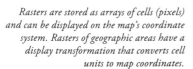

Rasters are stored as arrays of cells (pixels) and can be displayed on the map's coordinate system. Rasters of geographic areas have a display transformation that converts cell units to map coordinates.

In this simplistic case, six parameters define how a raster's rows and columns transform onto map coordinates. Often, a more advanced equation is applied to define the transformation from image space to a coordinate system.

Raster operators

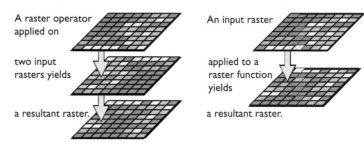

A raster operator applied on two input rasters yields a resultant raster.

An input raster applied to a raster function yields a resultant raster.

When you are studying an area, you may want to apply a suitability analysis. To do this, you would select rasters with values such as rainfall, soil alkalinity, and insolation and apply a series of operators according to your formula for suitability. Operators can be arithmetic, Boolean, relational, bitwise, combinatorial, logical, accumulative, and assignment.

Map calculation

Mathematical operations can be applied to two rasters and the result is in the output raster. Functions include +, –, /, *, Log, Exp, Sin, Cos, and Sqrt.

Map query

You can apply Boolean and logical operators on two rasters to create an output raster with true/false values. Operators include And, Or, XOr, Not, >, >=, =, <>, <, and <=.

Raster functions

There are many raster functions. Each can accept one or many rasters as input and generate one or several rasters with the calculated results.

Local functions perform a calculation on a single cell at a time. The neighboring cells do not influence the result. The functions can be applied to one raster or several overlaid rasters. Local functions include trigonometric, exponential, logarithmic, reclassification, selection, and statistical functions.

Focal functions perform a calculation on a single cell and its neighboring cells. A neighborhood can be a rectangle, circle, annulus (doughnut), or wedge. These functions can return the mean, standard deviation, sum, or range of values within the immediate or extended neighborhoods.

Zonal functions perform a calculation on a zone, which is a set of cells with a common value. The cells that form a zone can be discontinuous. There are two categories of zonal functions: statistical and geometric. These functions include area, centroid, perimeter, ranges, and sum calculations.

Global functions perform a computation on the raster as a whole. Examples are the calculation of Euclidean distances, weighted cost distances, and watershed delineation.

Images and other raster datasets are important GIS data resources that can be expensive to build and maintain. The decisions you make for storing image and raster data are influenced by how you will use the data.

A raster dataset can represent either a small area or can comprise a mosaic of a large number of input raster files.

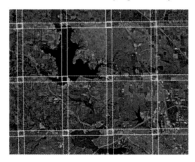

Raster catalogs are collections of raster datasets. When you draw a raster catalog, either individual raster datasets are displayed or their wireframe boundaries, depending on display settings that you control.

Mosaic datasets combine the best features of raster datasets and raster catalogs. Mosaic datasets dynamically fuse the display of input raster files with on-the-fly drawing methods such as pan-sharpening, display fast using high-performance network data access, and leave input raster files unaltered. It's easy to update mosaic datasets with new input raster files.

Raster data differs from vector data because it's not often modified and usually does not undergo continual edits. Raster data is generally stored in its original or similar form. It may be modified or improved upon for usability by changing the spatial reference (through the techniques of georeferencing or orthorectification) or by enhancing its appearance, but individual pixel values are not normally directly altered. Unlike vector data, which has point, line, and polygon shapes that are frequently edited, pixel values in raster data are not usually directly edited. New raster data is either directly added to a geodatabase or derived raster products are built from the source raster data and added to the geodatabase.

There are three methods to store raster data: as files in a file system, as rasters in a geodatabase, or as rasters managed from within the geodatabase but stored in a file system. Which method you choose is influenced by how the data will be used, with cost as another consideration.

Raster datasets

A raster dataset is any valid raster format that is organized into one or more bands. Each band consists of an array of pixels; each pixel has a value. Multiple raster datasets can be mosaicked into a larger, single, continuous raster dataset. Raster datasets can be stored inside any type of geodatabase (personal, file, or ArcSDE) or as files on disk. This gives you many options for storing the physical raster data.

Raster catalogs

A raster catalog is a collection of raster datasets. Raster catalogs manage raster datasets using a table structure wherein each row defines the individual raster datasets, and additional columns can be added to define attributes. A raster catalog is often used to display adjacent, fully overlapping, or partially overlapping raster datasets without the requirement to mosaic them into one large raster dataset. Raster catalogs can either fully contain raster datasets in the geodatabase (which are called managed raster catalogs) or they can reference raster datasets stored as files on disk (which are called unmanaged raster catalogs).

Mosaic datasets

The mosaic dataset is the optimum geodatabase data model to store and manage your image and raster data in ArcGIS. The mosaic dataset is a collection of raster datasets (images) stored as a catalog and viewed as a mosaicked image. These collections can be extremely large both in total file size and number of raster datasets.

A mosaic dataset is created in a geodatabase and can have raster datasets added to it directly, or it can be created from the whole or a selection set within a raster catalog or other mosaic dataset. Therefore, data stored using a managed or unmanaged raster catalog can be utilized and any additional attribute fields a user may have created in a raster catalog will be maintained. By allowing a mosaic dataset to contain a mosaic dataset, you can manage specific collections of data in one mosaic dataset, while providing a 'master' mosaic dataset to your data users. Or you can create a single mosaic

The mosaic dataset can most quickly access raster data if your organization uses technologies such as network access-storage (NAS) or storage-area network (SAN). These are computer architectures combining computer servers with arrays of disk drives for rapid data access.

A raster type has sophisticated capabilities and can recognize that a particular image, such as one provided by a satellite imaging company, includes a raster dataset with several bands, varying spatial resolutions, and other metadata that affect the spatial reference. Therefore, if the product had four bands of data at a 1 meter resolution, and one band with a 30 cm resolution, the raster type may know to create a product that sharpens the lower resolution data with the higher resolution dataset (also known as pan-sharpening). Additionally, if the correct rational polynomial coefficient (RPC) information is provided, this raster type could be used to improve on the fused data product by performing an orthorectification.

An example of applying raster functions with mosaic datasets is when specific raster data products (such as from a satellite sensor) are added to a mosaic dataset and some functions are automatically added to the raster data. As mentioned above, you could add raster datasets that are used to generate an orthorectified, pan-sharpened image. To generate this image, both a pan-sharpen function and orthorectification function would be applied to the raster data when it is accessed. This is advantageous because it saves disk space, since you aren't required to store source and preprocessed datasets. Additionally, if you wanted to process the same data differently, you can add the same data to a different mosaic definition and apply different functions. You may still want the orthorectification function used, but you may want to generate a vegetative index. You could use the Band Algebra function or the NDVI function.

dataset of your entire collection, and create specific mosaic datasets to provide collections to specific users.

A mosaic dataset manages its raster data in the same way as an unmanaged raster catalog; therefore, the tables will be similar, datasets will be indexed, and queries can be performed on the collections.

Mosaic datasets can manage data stored inside its geodatabase or another geodatabase, or on a file storage system such as a local disk or using a network device such as a NAS or SAN. Mosaic datasets can be faster to create than other data storage models because they don't move or alter the data files. Instead, when you create a mosaic dataset, indexes are created for raster datasets and the header and attribute information is read.

The raster data in a mosaic dataset does not have to be adjoining or overlapping but can exist as unconnected, discontinuous datasets. The data can even be completely or partially overlapping but be captured over different dates. The mosaic dataset is an ideal dataset for storing temporal data. You can query the mosaic dataset for the images you need based on time or dates and use a mosaic method to display the mosaicked image according to a time or date attribute.

Mosaic datasets and raster types

Mosaic datasets utilize raster types to read and ingest the required information from raster datasets. It identifies metadata, such as georeferencing, acquisition date, and sensor type, along with a raster format. A raster type can read raster data in its simplest way, by just using a raster format such as TIFF or JPEG. A raster type understands how to read and display the pixel data and apply the spatial reference associated with raster datasets. By using the correct raster type, you can automatically define functions that will be applied on-the-fly when the raster datasets are accessed.

Applying raster functions with mosaic datasets

Functions are operations that are applied on-the-fly to a raster dataset within the mosaic dataset or to all the contents of the mosaic dataset to deliver processed raster data to the user. You can add functions to raster dataset or mosaic dataset, or they may be added when the data is added to the mosaic dataset.

Managing multiple resolutions with mosaic datasets

Mosaic datasets are designed to handle data with varying resolutions—spectral, spatial, temporal, and radiometric. The raster types and functions in a mosaic definition play a strong role in how all of this data is handled and displayed. Additionally, the mosaic dataset is particularly aware of the spatial and temporal information as attributes of the raster data. Based on the cell sizes, the mosaic dataset will display the imagery at the most appropriate scales. With some additional display control properties, called mosaic methods, a user can control the temporal information allowing them to view the images for the dates they require.

You use mosaic methods in a mosaic dataset to control what raster data is presented each time a mosaic (from the mosaic dataset) is displayed. By default, the mosaic is generated by displaying the raster dataset that is the closest to the center of the image. Another mosaic method lets you define a query based on attributes such as acquisition date or cloud cover. These mosaic methods and querying capabilities allow users to have access to every raster dataset within the mosaic dataset, even when there is overlap. You can also query a mosaic dataset based on your spatial and nonspatial query constraints. The results of that query can limit what images are displayed by the mosaic dataset, or you can display each one as an individual image layer.

There is no pixel data loss or metadata loss when using a mosaic dataset, as the source pixels are never altered or converted, and the files are never moved, therefore, any metadata files remain in their location. Because the mosaic dataset does not alter the source data or its location, the pixel values are not altered. Additionally, the mosaicking performed by the mosaic dataset occurs on-the-fly only when the mosaic dataset is accessed. Users have access to the mosaicked image as well as the source data; therefore, there is no data loss occurring for overlapping datasets.

Advantages of mosaic datasets

Mosaic datasets are excellent data models for storing and managing data. Mosaic datasets are ideal for distributing data because they can be directly accessed by users and easily served.

	Raster dataset	Mosaic dataset	Raster catalog
Description	A single picture of an object or a seamless image covering a spatially continuous area. This may be a single original image or the result of many images appended together.	A collection of raster datasets stored as a catalog that allows you to store, manage, view, and query collections of raster data. It is viewed as a mosaicked image, but you have access to each raster dataset in the collection.	A collection of raster datasets displayed as a single layer. They can be in different coordinate systems and can have different data types. It is viewed as a wireframe by default.
Storage	Stored in a geodatabase or in a file system.	Stored in a geodatabase or in a file system (only in special cases). Raster data can be stored within the geodatabase or outside.	Raster data can be stored within the geodatabase or outside.
Size limit	Size limit is dependent on how the dataset is stored.	Each raster dataset within the mosaic dataset has the same size limit as a single raster dataset. Limited by size of table—in a personal geodatabase, two gigabytes; in a file geodatabase, one terabyte; in a database accessed via ArcSDE, no practical size limit, as it is determined by the relational database management system. The limit is not affected by the size of each raster data, but generally by the number of rasters within.	Each raster dataset within the raster catalog has the same size limit as a single raster dataset. Limited by size of table—in a personal geodatabase, two gigabytes; in a file geodatabase, one terabyte; in a database accessed via ArcSDE, no practical size limit, as it is determined by the relational database management system. The limit is not affected by the size of each raster data, but generally by the number of rasters within.

	Raster dataset	Mosaic dataset	Raster catalog
Map layers	One map layer.	One map layer.	One map layer.
Homogenous or heterogenous data	Homogeneous data: a single format, data type, and file.	Heterogeneous data: multiple formats, data types, file sizes, and coordinate systems.	Heterogeneous data: multiple formats, data types, file sizes, and coordinate systems.
Metadata	Stored once and applies to complete dataset.	Can be stored within the raster dataset item in the mosaic dataset as well as attributes in the raster catalog table and can be on the mosaic dataset.	Stored as attribute columns for each raster dataset item in the raster catalog; metadata is also stored for the raster catalog.
Pyramids or overviews	A single pyramid on the entire raster dataset.	Pyramids for each raster dataset, as well as overviews (like a pyramid) for the entire collection.	A pyramid for each raster dataset in the raster catalog.
Mosaicking	When multiple raster datasets are mosaicked into one raster dataset, this results in a loss of data at overlapping areas.	Mosaicking on mosaic datasets is done on-the-fly and controlled by user, therefore, there is no loss of data.	No mosaicking occurs in a raster catalog; therefore, there is no loss of data.
Pros	• Fast to display at any scale. • A mosaicked raster dataset saves space, since there is no overlapping data.	• Manages large collections of raster data. • Fast to display at any scale. • No loss of data to create mosaic. • User has access to full content of collection. • Properties can be set to control the mosaicked display. • On-the-fly processing.	• Can manage multi-row raster tables for many purposes. • Can specify one or more raster datasets for display. • Can view large raster catalogs as a wireframe view with a polygon representing the geometry for each raster dataset within the raster catalog.
Cons	File and personal geodatabase raster datasets are slower to update because the entire file has to be rewritten.	Overviews can take time to generate.	File-based raster catalogs with different data types may not render well. Displaying many raster dataset items from a raster catalog can be time consuming.
Serving	Can be served directly as an image service.	Can be served directly as an image service.	Can be served as an image service by first creating a referenced mosaic dataset that references the raster catalog.
Recommendations	Use a mosaicked raster dataset when overlaps between mosaicked images do not need to be retained and for fast display of large quantities of raster data.	Use a mosaic dataset for managing and visualizing raster data. Good for multidimensional data, for querying, storing metadata, and for overlapping data. Provides a good hybrid solution.	Use a raster catalog for massive image repositories, retaining overlaps between datasets, managing time series data, and when differences among adjoining images prevent mosaicking.

Designing and implementing a GIS using raster data is no different than it would be for any other GIS. The only difference is now you will be using raster data instead of, or in addition to, vector feature data.

The workflow steps for designing and building a geodatabase of raster data are straightforward. These seven steps outline a general methodology that can be adapted to a wide variety of projects.

When working with raster data, your workflow could be the following:

1 Identify purpose or objective.

Why do you need raster data? How do you wish to use raster data in the GIS? Uses generally fall in either or both of two categories: data for analysis and data for display. For example, raster data for analysis may involve a watershed analysis or terrain analysis, updating some topographic features in other datasets, or updating land-cover classes to assess the new location of a housing development. An example of data for display is the common use of orthophotos as a background for a map.

2 Identify the data.

If you're looking to extract information from imagery, consider the resolution you require and whether you need one or more spectral bands. You may consider whether the data can come from an aircraft or satellite. If you're going to work with elevation data, you may consider the most appropriate methods for collection, such as lidar, contour lines, or radar interferometry. For a collection of scanned maps, you need to identify what those maps are, such as scanned documents, CAD drawings, or topographic maps.

3 Refine the requirements.

Determine more detailed requirements based on:

- Cost—What are your budget limits? Can you afford the data you want? Is there an alternative within your budget?

- Availability—Does the data already exist? How often is the data updated? Will you receive updates as individual tiles or a single update with complete coverage? Can you receive this data in a timely manner?

- Licenses—Can you share or distribute this data? Can you use this data in multiple projects? What can you do with the information or data derived from the original data? Can you serve this to the public using the Internet?

- Resolution—Will the available level of detail provide the required information?

- Storage—What database or file formats will be used? How large is each file? Will you use pyramids? How much total disk space is needed?

- Extent—Can you cover the area of interest with one raster image, or will you need multiple raster datasets?

- Accuracy—Will the available data resolution provide you with the required spatial accuracy? What is the level of accuracy promoted by the data vendor? How will the data be verified and validated?

- Accessibility and pricing—Is or will the data be accessible on a network? Will you charge fees for usage or downloads? Who will have access to the data? How will you control access and sales?

4 Acquire and review data.

This can involve placing orders for the data with a company capable of providing it, scanning the maps you need, or acquiring the source data and building the corresponding raster datasets. It is important that you have a system for checking the quality of the data, whether created in-house or acquired from outside sources. You may have to check for missing data (such as dropped lines or pixels), poorly represented data, or if the data is georeferenced for your area of interest.

5 Prepare the data.

Building the database could require the prior extraction or conversion from one data format to another, such as from lidar elevation points to a DEM. It could also involve some preprocessing, such as georeferencing or rubbersheeting.

6 Design and build the database.

There are several recommended choices:

• Build a mosaic dataset to manage all the imagery.

• Retain the data in separate image files.

• Build a large, seamless raster dataset (mosaic) from multiple images.

Additional considerations include which compression method to use and whether to use a personal geodatabase or a multiuser geodatabase management system.

You will need to create some level of metadata, depending on your intended distribution and access to the data. For example, what kinds of queries should users expect to use to find your raster data over the Web? If using mosaic datasets or raster catalogs, you may consider additional catalog fields to allow more extensive querying capabilities.

7 Deploy and maintain the geodatabase.

There are two main ways to provide access to the data, either by providing access to the data in the geodatabase or by serving the data. Raster datasets and mosaic datasets can be served as image services. Raster catalogs cannot be served directly as image services. You must create a mosaic dataset, add the raster catalog to it, and then serve the mosaic dataset. You can serve other resources that contain these data models, such as a map document. A map document can contain each of these three data models; however, when served, the properties of the data's layer in the map service cannot be altered by the user.

You will have to plan for updates, modifications, and the ability to build on your initial implementation. Mosaic datasets provide robust methods for updating. For example, when adding new data to a mosaic dataset it can still be accessed by users while new data is being added, therefore, there is no downtime for that data source. Additionally, if you replace the source files they are just read directly as long as their properties (such as size or bands) haven't changed. Or if you add new files to the source directory, these can be easily detected and added to the mosaic dataset using a synchronize process.

Surface modeling with terrains

8

The surface of the Earth is modeled in a geodatabase with terrain datasets. Terrain datasets can handle many millions and even billions of points with x,y,z values and produce realistic scenes of a landscape. Terrain datasets are well suited for careful shaping of features such as roads and rivers. From a terrain dataset, you can export an elevation raster. With this elevation raster, you can perform surface analysis such as stream and watershed delineation and improved cartographic display such as elevation gradient and hillshading combined, as shown on this map.

*The TIN representation models a surface from a set of points from which triangles are formed, or **triangulated**.*

*Triangles are made from three points that occur at **irregular** locations.*

*Each triangle stores topological information about its neighboring triangles, thus forming a **network**.*

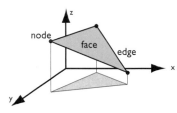

Three points in space uniquely define a face, a triangular section of an infinite plane. Faces join edge to edge in a triangulation and each face has a slope (vertical angle) and aspect (horizontal direction).

A constrained Delaunay triangulation is the same as Delaunay with the exception of breaklines, those linear features enforced in the triangulation as triangle edges. A Delaunay conforming triangulation often needs to densify vertices of breaklines to ensure the resulting triangles honor the Delaunay point in circle test. Constrained triangulations do not try to enforce the Delaunay rule along breaklines, so the lines are not densified.

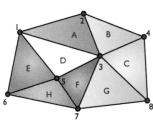

Triangle	Node list	Neighbors
A	1, 2, 3	–, B, D
B	2, 4, 3	–, C, A
C	4, 8, 3	–, G, B
D	1, 3, 5	A, F, E
E	1, 5, 6	D, H, –
F	3, 7, 5	G, H, D
G	3, 8, 7	C, –, F
H	5, 7, 6	F, –, E

A simple triangulation consisting of eight triangles and eight nodes. A TIN stores topological associations among triangles.

Terrain datasets are based on TINs (triangulated irregular network), which have been used by the GIS community for many years and are a digital means to represent surface morphology. TINs are a form of vector-based digital geographic data and are constructed by triangulating a set of points. The vertices are connected with a series of edges to form a network of triangles. There are different methods to form these triangles, such as Delaunay triangulation or minimum slope. ArcGIS supports the Delaunay triangulation and constrained Delaunay triangulation methods.

This graphic shows the criterion for the Delaunay triangulation for the simple case of four points building two triangles.

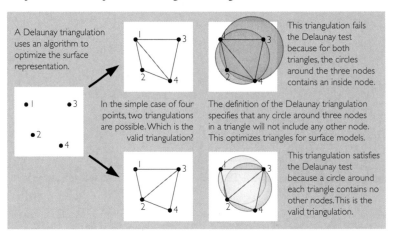

A Delaunay triangulation uses an algorithm to optimize the surface representation.

In the simple case of four points, two triangulations are possible. Which is the valid triangulation?

This triangulation fails the Delaunay test because for both triangles, the circles around the three nodes contains an inside node.

The definition of the Delaunay triangulation specifies that any circle around three nodes in a triangle will not include any other node. This optimizes triangles for surface models.

This triangulation satisfies the Delaunay test because a circle around each triangle contains no other nodes. This is the valid triangulation.

The second triangulation satisfies the Delaunay triangle criterion, which ensures that no vertex lies within the interior of any of the circumcircles of the triangles in the network. If the Delaunay criterion is satisfied everywhere on the TIN, the minimum interior angle of all triangles is maximized. The result is that long, thin triangles are avoided as much as possible.

Topology of a TIN

The topological structure of a TIN is defined by maintaining information defining each triangle's nodes, edge numbers, type, and adjacency to other triangles. For each triangle, a TIN records:

- The triangle number
- The numbers of each adjacent triangle
- The three nodes defining the triangle
- The x,y coordinates of each node
- The surface z-value of each node
- The edge type of each triangle edge (hard or soft)

In addition, the TIN maintains a list of all the edges that form the TIN's hull and information defining the TIN's projection and units of measure.

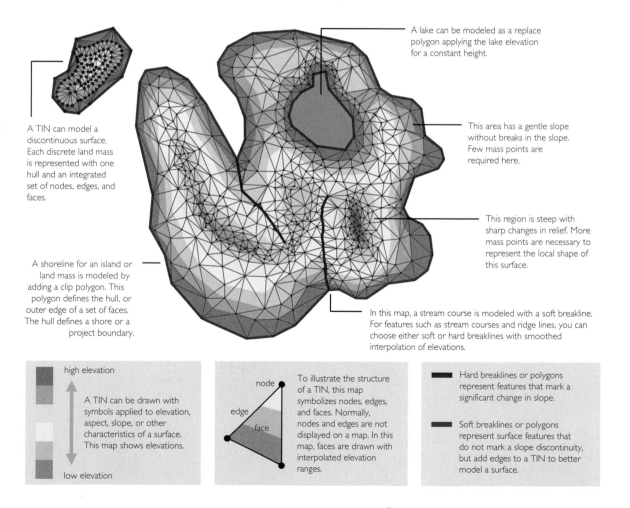

These are some common surface features found on topographic maps. They can be used to precisely define a TIN in a local region and they can also be derived from a TIN.

Surface features in a TIN

The edges of TINs form contiguous, nonoverlapping triangular facets and can be used to capture the position of linear features that play an important role in a surface, such as ridgelines or stream courses. Because nodes can be placed irregularly over a surface, TINs can be sampled more densely in areas where a surface is highly variable or where more detail is desired and a lower sample density in areas that are less variable.

The input features used to create a TIN remain in the same position as the nodes or edges in the TIN. This allows a TIN to preserve all the precision of the input data while simultaneously modeling the values between known points. You can include precisely located features on a surface—such as mountain peaks, roads, and streams—by using them as input features to the TIN nodes.

A lake can be modeled as a replace polygon applying the lake elevation for a constant height.

This area has a gentle slope without breaks in the slope. Few mass points are required here.

A TIN can model a discontinuous surface. Each discrete land mass is represented with one hull and an integrated set of nodes, edges, and faces.

This region is steep with sharp changes in relief. More mass points are necessary to represent the local shape of this surface.

A shoreline for an island or land mass is modeled by adding a clip polygon. This polygon defines the hull, or outer edge of a set of faces. The hull defines a shore or a project boundary.

In this map, a stream course is modeled with a soft breakline. For features such as stream courses and ridge lines, you can choose either soft or hard breaklines with smoothed interpolation of elevations.

high elevation

A TIN can be drawn with symbols applied to elevation, aspect, slope, or other characteristics of a surface. This map shows elevations.

low elevation

To illustrate the structure of a TIN, this map symbolizes nodes, edges, and faces. Normally, nodes and edges are not displayed on a map. In this map, faces are drawn with interpolated elevation ranges.

Hard breaklines or polygons represent features that mark a significant change in slope.

Soft breaklines or polygons represent surface features that do not mark a slope discontinuity, but add edges to a TIN to better model a surface.

One of the most important layers in any GIS is a layer that accurately portrays the terrain of the Earth's surface. With terrains, we can make three-dimensional perspectives and do surface analysis, such as visibility areas. Studying water flow from a cloudburst to the drainage network requires a detailed surface model.

Terrain datasets

1:24,000 detail of terrain display

A terrain dataset is a multiresolution, TIN-based surface built from measurements stored as features in a geodatabase. Terrain datasets are made from data sources such as stereo-captured photogrammetric features and mass point collections of 3D data such as lidar, sonar, and bathymetry.

Terrains reside in the geodatabase, inside feature datasets with the features used to construct them. The feature dataset contains feature classes with data sources for the terrain dataset.

Terrain datasets in the geodatabase are designed to handle voluminous sets of point data from many sources. Terrain datasets can also integrate other 3D data, such as lake boundaries and breaklines for edges of roads, that are collected with GPS receivers and other methods.

1:12,000 detail of terrain display

There is increasing use of lidar and other sensors for collecting high-resolution and massively large point datasets of elevation observations. Geodatabases are designed for managing these critical data assets as well as for integrating these and other data sources into integrated terrain datasets.

Many users want to:

- Better represent and model the terrain of their study areas by integrating their 3D-based mass point observations with other data sources such as 3D features captured using stereo photogrammetry.

- Use terrains for many types of 3D spatial analysis in their GIS using the ArcGIS 3D Analyst extension.

- Derive raster-based digital elevation models for use in modeling and analysis systems such as the ArcGIS Spatial Analyst extension.

1:6,000 detail of terrain display

A terrain dataset is a TIN-based data structure with multiple levels of resolution. Terrain datasets include pyramids that provide the appropriate levels of detail for use at multiple scales.

A terrain dataset is built from multiple data sources such as lidar mass point collections, 3D breaklines, and 3D-based survey observations. The data sources used to create terrain datasets are managed as a set of integrated feature classes in the geodatabase.

Because terrain datasets are stored in a feature dataset, they can be versioned. This allows for multiuser editing environments.

You can find many surface models in raster form, but these are often derived from a triangulated irregular network of elevation points. You can export a terrain dataset to an elevation raster and use the many analytic and cartographic tools that work on elevation rasters.

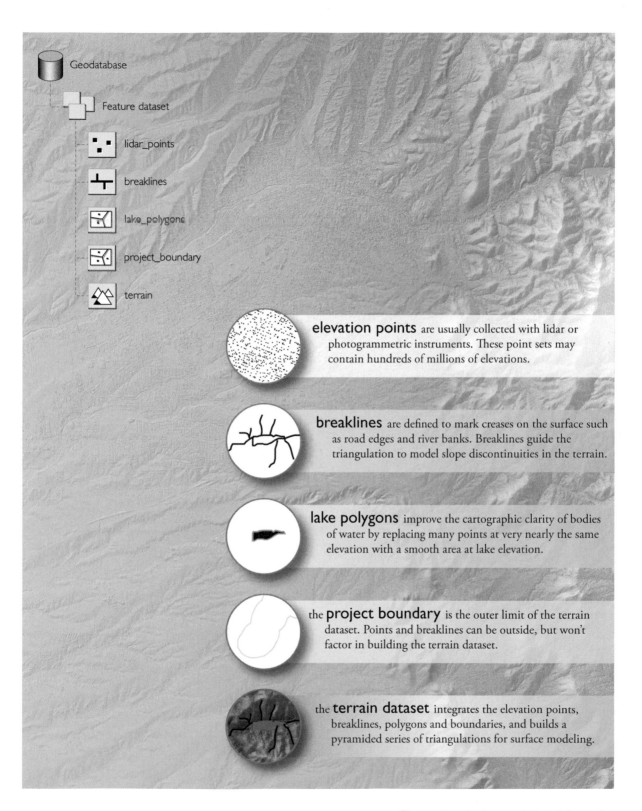

Geodatabase

Feature dataset

lidar_points

breaklines

lake_polygons

project_boundary

terrain

elevation points are usually collected with lidar or photogrammetric instruments. These point sets may contain hundreds of millions of elevations.

breaklines are defined to mark creases on the surface such as road edges and river banks. Breaklines guide the triangulation to model slope discontinuities in the terrain.

lake polygons improve the cartographic clarity of bodies of water by replacing many points at very nearly the same elevation with a smooth area at lake elevation.

the **project boundary** is the outer limit of the terrain dataset. Points and breaklines can be outside, but won't factor in building the terrain dataset.

the **terrain dataset** integrates the elevation points, breaklines, polygons and boundaries, and builds a pyramided series of triangulations for surface modeling.

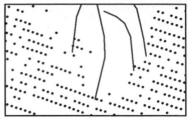

Lidar points with breaklines for ridges along a mountain top.

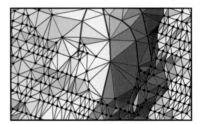

Triangulation on points and breaklines.

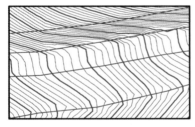

Elevation gradient, hillshading, and triangulation.

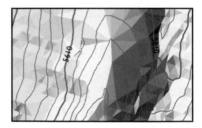

Elevation gradient, hillshading, and calculated 2-foot contours. Map scale is 1:400; successive lidar points are spaced about one foot apart.

Terrain datasets can be made from different types of data. These include lidar and sonar points, breaklines and points derived from stereo photography, and other forms of survey data. The supported geometry types for input feature classes include points, multipoints, polylines, and polygons. Multipoints are used for large datasets such as lidar point collection.

The ability to incorporate a variety of data types into the definition of a surface offers maximum control to sculpt an accurate representation. Surface-specific points capture peaks and pits. Mass points add overall form and control. Breaklines indicate abrupt changes in slope that occur across linear features. Polygons delineate flat areas or areas of no data.

Hard or soft surface feature types

Hard and soft qualifiers for line and polygon feature types are used to indicate whether a distinct break in slope occurs on the surface at their location. This information influences the behavior of the natural neighbors interpolator. It interprets the terrain surface as smooth except when crossing hard lines and hard polygon boundaries. The natural neighbors interpolator is offered by tools that export a terrain to a raster, interpolate shape, and surface spot geoprocessing tools. All the surface feature types other than mass points support the hard or soft qualification.

Below are contours derived from a raster exported from a terrain dataset using the natural neighbors interpolator with soft and hard breaklines.

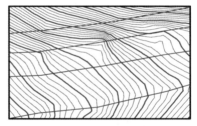

Contours derived from soft breaklines.

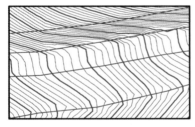

Contours derived from hard breaklines.

Some examples of hard features are lake shorelines, streams, building pads, curb lines along roads, and road cuts. Some examples of soft features are study area boundaries, ridge and valley lines for smooth or rolling topography, void area boundaries, and contours (contours can also be added as mass points).

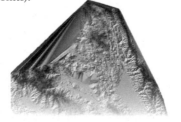

A terrain dataset mistakenly built without a clip polygon.

Same terrain dataset with a clip polygon for its boundary.

Point feature class

 Line feature class

 Polygon feature class

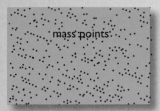

Mass points are used as nodes within the TIN. Many new sensors, such as lidar, can produce huge arrays of mass points that can be used to derive high-resolution terrain datasets.

Mass points are most frequently built from points. Contours are an example of line features whose vertices can be used as mass points. Polygon input is rare but available.

 Line feature class

 Polygon feature class

Breaklines are lines with height (z) recorded at each vertex. They become sequences of one or more triangle edges. Breaklines typically represent either natural features, such as ridgelines or streams, or built features, such as roadways.

 Polygon feature class

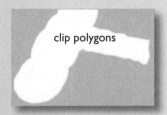

Clip polygons are used to define boundaries for terrain surfaces. They're needed when a data area has an irregular shape. Without a clip polygon, the data area will be convex.

 Polygon feature class

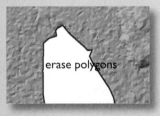

Erase polygons define holes in a terrain. These are used to represent areas for which you have no data or want no interpolation to occur. They will display as voids, and analysis will consider them to be areas of no data.

 Polygon feature class

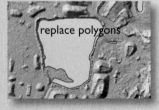

Replace polygons define areas of constant height. These are typically used to represent water bodies or man-made features that are flat. Replace polygons are best used when other measurements may exist in their interior that have different heights, and you want them reset.

 Polygon feature class

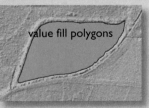

Value fill polygons are used to assign integers to the triangles of the terrain dataset when it's displayed or converted to a TIN. The boundary of each value fill polygon is enforced in the triangulation as a breakline, then the triangles within the polygon are assigned, or tagged, its value. These values are used to represent any user-defined criteria that can be encoded as an integer, such as land-use and soil type codes.

While mass points and breaklines compiled by photogrammetrists is still an important data source for terrains, the emerging dominant technology for terrain creation is lidar. The major advantage of lidar is the high volume of very accurate elevation points that are collected quickly and efficiently.

For geographic applications, a lidar system is mounted within a fixed-wing aircraft or helicopter. The lidar system is flown along a series of flight lines within a project area collecting laser elevation values from the surface below. Lidar (LIght Detection And Ranging) uses pulses of laser light to determine the distance from the laser system to the point of reflectance on the Earth's surface. A lidar system utilizes a rapidly rotating mirror to accurately position the high volume of laser points. The elevation points are accurately positioned using an INS system (Inertial Navigation System), which contains a GPS (Global Positioning System) and an IMU (Inertial Measurement Unit).

A lidar survey can collect many millions or billions of elevation points creating massive amounts of data to be manipulated and stored. The terrain dataset is engineered to handle the huge data volumes produced by lidar. Terrain datasets scale well by using a strategy employed by raster datasets—pyramiding, which is the progressive thinning of data at a series of map scales.

The surface resolution produced by lidar technology is unprecedented. Features such as buildings and trees are easily discernible and many new avenues of surface analysis are possible. For example, tree canopies can be mapped and measured and floodplain analysis made more accurate. A detailed view of the surface can reveal archaeological sites and seismic fault lines. In India, a cadastre has been built using parcel lines from hedge rows found with lidar.

Importing lidar data

Lidar data comes from the lidar provider in several forms.

The standard industry format for exchanging lidar data is the LAS format. LAS is the file extension of a binary file with lidar points with their attributes. This standard is defined by the ASPRS (American Society for Photogrammetry and Remote Sensing) and is an open format containing information for each point, including the number of returns, intensity, x,y,z positional values, scan direction, GPS time, and optional classification codes. Lidar data is also commonly processed into text files commonly containing x, y, and z positional values for elevation points.

When you import massive point datasets of lidar, it is important to write the output to a multipoint (instead of point) feature class. The high volume of elevation points in lidar data makes the use of multipoints necessary. Many lidar points are stored within one multipoint feature, and this reduces the number of rows in the feature class table, which improves overall performance. The 3D Analyst extension to ArcGIS provides import tools for lidar data delivered in either LAS or text file formats. Use the 3D Analyst import tools to create multipoint feature classes for lidar data.

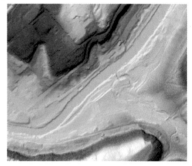

The bare earth surface drawn with hillshading and elevation color ramp.

The full surface with trees and buildings visible, drawn with hillshading and elevation color ramp.

First return and ground points

The lidar provider can supply the full or bare earth surface point sets, or if you have access to the LAS files, you can derive these full surface or bare earth surface point sets yourself.

Lidar data models not only the ground-level surface, but every sizable tree, building, or other protruding object. With lidar data, you can build terrain datasets for the full surface with trees and buildings or the bare earth surface with these objects removed. Some applications, such as forestry and urban planning, are interested in defining objects by the difference of these two surfaces. Both types of surfaces, full and bare earth, have many applications and you'll probably want both terrain datasets in your geodatabase.

The map with raw lidar points contains all the points that were collected.

The bare earth lidar points have buildings, trees, and very dense areas filtered out.

Raw lidar points   *Lidar filtered to bare earth points*

The raw lidar data, when drawn at large map scales, shows structures and trees. The bare earth map shows the surface as if these objects were removed.

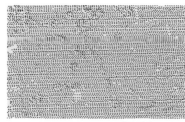

Raw lidar with hillshading and graduated colors for elevation.   *Bare earth lidar with hillshading and graduated colors for elevation.*

Map display and contours from lidar data

A characteristic of lidar data is its highly detailed model of the surface, with many undulations and small spiked objects. While lidar produces an accurate and detailed representation of a surface, it is too detailed for some applications. Drawing a terrain dataset layer on a map generally gives a good rendering, but at large scales, noise from the limits of vertical accuracy as well as rocks and shrubs in the elevation data becomes distracting. Present day lidar technology has a vertical accuracy of 12 to 15 cm, which becomes significant when average point spacing is a meter or less.

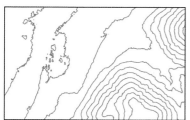

Contours made from a full resolution terrain dataset are accurate, but too detailed for some maps.

To produce smooth contours from a terrain dataset, a recommended practice is to first export to an elevation raster dataset using less than the full resolution pyramid level of the terrain dataset. Using a slightly filtered point set will filter out much of the noise in vertical measurements. Smooth contours can then be made by using nearest-neighbor interpolation for the raster export from the terrain dataset. While slightly less accurate, elevation rasters generate smoother contours like those on topographic maps.

Contours made from an elevation raster exported from a terrain dataset are smoother and look better on a topographic map.

With this output elevation raster, you can also use many analytic tools for functions such as delineating streams and for producing better hillshading of a surface.

When you create a terrain dataset, these are the properties you set.

Average point spacing

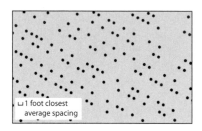

Bare earth lidar points show a pattern of points spaced along lines. With this data, closest points are spaced one foot apart.

When a terrain dataset is created, it needs to be told the average point spacing of the input measurements. The terrain dataset uses this information to define a horizontal tiling system into which input measurements are divided. The average point spacing is used as a means to bin, or group, the points together constructing a virtual tile system.

This tile system improves the performance of spatial queries. It also helps split the data up into manageable chunks. For the most part, the tile system is internal and managed by a terrain dataset. Otherwise, you would, particularly in large projects, need to handle this in some well-orchestrated set of specifications and data processing steps.

Usually, the average point spacing is defined as part of the data procurement process and is recorded as metadata. If you don't know the average point spacing of your data, you'll need to determine what it is. The best average spacing to use for terrain is that which represents the most common distance between points and vertices. For example, there may be some points very close, for example, 0.2 meters, and some very far apart, for example, 5 meters, but if the vast majority are approximately 2 meters, that is the value you should specify. Outliers should have little to no influence.

Height source

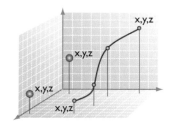

Points, multipoints, polylines, and polygons have optional z-values.

When adding a feature class to a terrain dataset, you need to indicate whether or not it has z-values and, if so, where they come from. In the case of 3D features, the z-values reside with the shape geometry. Indicate the Shape field as the source because this is a reference to the geometry. You can tell whether or not a feature class is 3D by reviewing its properties in the catalog. Alternately, look at the Shape field in a table view of the feature class. If the listed geometry type includes a Z, such as PolylineZ and PolygonZ, it is 3D.

You can also use 2D features with z-values stored in an attribute field. It's not uncommon for points to be stored this way. The limitation for 3D geometry types other than points is that each feature must be flat (that is, have a constant height); since there's only one z-value to use for all the vertices. Contours and water bodies are examples of features that can be stored this way. The height source for a 2D feature class like this is the name of the attribute field containing the z-values.

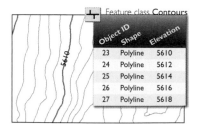

Certain flat objects, such as lake shores and contours, can have elevation as an attribute.

While features used to create a terrain dataset normally have z-values from either their geometry or an attribute, there are exceptions. Clip polygons are the most common of these. They're needed to properly delineate the data area of the surface but are hard to obtain in 3D. The terrain dataset supports the inclusion of 2D features by first creating an intermediate surface from all the 3D features and interpolating the 2D feature's heights on this surface. This converts them, in memory, into 3D features that are then incorporated into the surface.

Surface feature types in a terrain	
✓ Lidar points	Mass points
✓ Ridge lines	Soft breaklines
✓ Road cuts	Hard breaklines
✓ Lakes	Replace polygons
✓ Boundary	Clip polygons

A simple terrain combining lidar data with photogrammetric breaklines, represented with five common surface feature types.

Same extent shown at three pyramid levels.

Overview is the coarsest pyramid level, scaled for good display at the full terrain extent.

Pyramid level	1	2	3	4
Stream lines	✓			
Road cuts	✓	✓		
Lakes	✓	✓	✓	
Boundary	✓	✓	✓	✓

You can control which surface feature types participate at each pyramid level.

Surface feature type

The feature class will play in defining the terrain dataset surface. There are mass points, breaklines, and several polygon types. Breaklines and polygons also have hard and soft qualifiers. These indicate to an interpolator whether the surface crosses over the features smoothly (soft) or with a potentially sharp discontinuity (hard). Point and multipoint feature classes can only be represented as mass points. Polyline and polygon feature classes can be represented as breaklines, clip polygons, erase polygons, replace polygons, and value fill polygons

Terrain pyramid type

Pyramids are levels of detail generated for a terrain dataset to improve efficiency for some applications. They are used as a form of scale-dependent generalization. Pyramid levels take advantage of the fact that accuracy requirements diminish with scale. They are similar in concept and purpose to raster pyramids, but their implementation is different.

Overview terrain

The overview terrain is the coarsest representation of the terrain dataset and is intended for fast drawing at small scales. The overview is what's drawn by default when zoomed to the full extent of the terrain dataset. It is often referred to as a vector-based thumbnail representation.

Groups

Groups are used to define multiple levels of detail for line and polygon features. Since terrain datasets don't have an automated way of generalizing polylines and polygons, you need to specify how a terrain dataset uses features through group definition. An example is creating a detailed clip polygon feature class that should only be used as large scales and a generalized version for use at small scales, such as edge of pavement polygons for large scales and road centerlines for small scales.

Embedded feature classes

Terrain pyramids can require a significant amount of storage space. The size is roughly equivalent to that of the geometry present in the feature classes participating in the terrain. For large point collections, typically represented by lidar or sonar, the cost may be prohibitive. In these cases, large multipoint feature classes can be embedded in a terrain dataset.

Pyramid resolution bounds

Minimum and maximum resolution bounds are used for feature classes added as polyline or polygon surface feature types. They define the range of pyramid levels the features will be enforced in the surface. You provide the resolution thresholds given in the resolution of the terrain's pyramid levels.

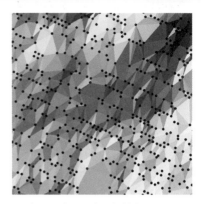

A large scale map detail of lidar points on a shaded terrain dataset. The space between closest points is about one foot (.3 meter).

Z tolerance terrain pyramid

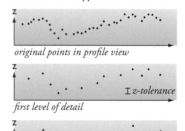

original points in profile view

first level of detail

second level of detail

Window size terrain pyramid

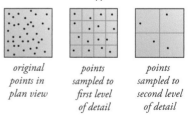

original points in plan view

points sampled to first level of detail

points sampled to second level of detail

Terrain pyramiding improves performance by providing a scale-dependent means of data reduction. Pyramids reference only the data needed to construct a surface of an approximate accuracy. On-the-fly surface construction, display, and analysis are faster for smaller-scale applications because only a thinned subset of the data is required. The original data is not moved or averaged in any way. The exact positional information of the measurements is maintained.

With modern data collection technology, we can build extremely detailed terrain models using millions to billions of elevation points. These are the most common data collection technologies for terrain datasets.

- Lidar is a new technology providing large datasets from laser pulses reflected from the ground during an aerial lidar survey.

- Photogrammetric data collection involves a technician operating a stereo compiler and collecting points and detailed breaklines.

- GPS data collection can also be used to enhance terrain models by adding road and trail lines and points on the surface. Survey-grade GPS receivers using differential correction can locate positions with a centimeter level accuracy. Hand-held GPS receivers are not accurate enough for augmenting most lidar datasets; they have typical accuracies of 5 to 10 meters, larger then most lidar average point spacings.

Scalability and terrain datasets

Terrain datasets are designed to handle vast amounts of elevation data and diverse surface feature types by using terrain pyramids, triangulations built on a progressively thinned sampling of data points. This is conceptually similar to raster pyramids, which are also subsamplings of a large dataset for efficient display and analysis at a full range of map scales.

Another way to achieve scalability is to use the multipoint shape type, which minimizes rows in a table, allowing more efficient and rapid data access.

Two types of pyramids

Two types of pyramids can be used to build a terrain dataset: z-tolerance and window size.

Using the z-tolerance pyramid type, pyramiding is accomplished through the application of a z-tolerance-based filter that is used to thin points. You eliminate noncritical points to produce derivate surfaces that are within an approximate vertical accuracy relative to the full-resolution data.

With the window size pyramid type, pyramiding is carried out through the designation of a window size filter. It thins points for each pyramid level by partitioning the data into equal horizontal sample density areas (windows) and selecting just one or two points from each area as representatives. Selection is based on one of the following criteria: the minimum, maximum, mean, or both the minimum and maximum z-value.

Breakline enforcement in pyramids

Additionally, the enforcement of lines and polygons is controlled on a per-pyramid-level basis. For example, breakline enforcement can be restricted to the highest one or two resolution pyramid levels. Some features, such as study area boundaries and lake shorelines, might need representation through all scales but not at the same detail. Generalized representations can be used at coarse scales, while full detail is only applied at larger scales.

These terrain details are displayed at progressing map scales and show the nominal pyramid levels for each map scale.

These terrain details are displayed at a fixed map scale of 1:12,000 and show progressing pyramid levels.

These triangulation maps are also displayed at a fixed map scale of 1:12,000 and show the triangle edges for progressing pyramid levels.

Overview *This pyramid level has no maximum scale and displays from 1:50,000*

Map scale of this terrain detail is 1:60,000

Overview

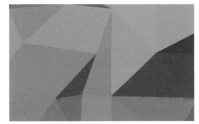

Map scale of this terrain detail is 1:12,000

Overview

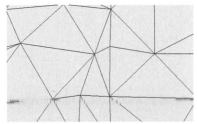

Map scale of this triangulation is 1:12,000

Window size 128 *This pyramid level displays from 1:20,000 to 1:50,000*

Map scale of this terrain detail is 1:40,000

Window size 128

Map scale of this terrain detail is 1:12,000

Window size 128

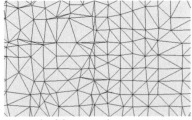

Map scale of this triangulation is 1:12,000

Window size 64 *This pyramid level displays from 1:10,000 to 1:20,000*

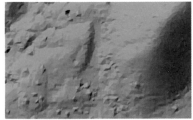

Map scale of this terrain detail is 1:15,000

Window size 64

Map scale of this terrain detail is 1:12,000

Window size 64

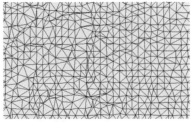

Map scale of this triangulation is 1:12,000

Window size 32 *This pyramid level displays from 1:5,000 to 1:10,000*

Map scale of this terrain detail is 1:6,000

Window size 32

Map scale of this terrain detail is 1:12,000

Window size 32

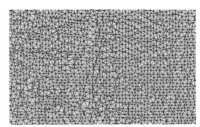

Map scale of this triangulation is 1:12,000

Contour lines are widely used in new and old maps. Contours present the advantage of letting you find an elevation directly on the map.

Visualization of surface elevation data is very important. Being able to view data in 3D, color coded with elevation, sun shaded, blended with attribute data, and even draped with imagery is becoming more customary. Three main regions exist for visualizing and working with elevation datasets: 2D map views, 3D scenes, and 3D globe views. ArcGIS provides applications that enable large elevation datasets to be stored and visualized using each distinct view. Rasters, TINs, and terrain datasets all provide a method to store and visualize terrain data in 3D.

Terrain datasets can be visualized as a 2D map in ArcMap and as a 3D globe in ArcGlobe through a terrain layer. This layer type is similar to TIN layers in some regards. It supports multiple renderers. You can view the triangles colored by elevation range, slope, aspect, and hillshade. You can also see the breaklines, triangle edges, and nodes of the triangulated surface. In terms of differences, terrains have the level of detail (LOD) capability that helps speed up the display, particularly at small scales, when a large volume of data is involved.

Surfaces can be displayed with a terrain layer or raster layers built from the terrain dataset. Use terrain layers for quick and accurate surface displays and use elevation rasters and derived analytic rasters for superior cartography and additional analysis options.

Drawing surfaces in two dimensions

Surfaces can be drawn in two-dimensions using techniques such as contour lines, elevation color gradients, and hillshading, which simulates the shaded relief of landscape.

With terrain datasets, you can draw the surface with techniques such as color ramps for elevations, hillshading for realistic relief presentation, and contours for lines of constant elevations. For diagnostic or special applications, you can also draw mass points, breaklines, polygons, and triangulations.

Drawing surfaces from rasters has several advantages. First, when you export a terrain dataset to a raster using a pyramid level of detail lower than full resolution and with nearest-neighbor interpolation, you can make a smoothed surface, free of small features which distract from the bare surface. Second, a rich set of raster tools is available to perform surface analysis, such as finding streamlines. Raster operations can be combined into geoprocessing models, as shown on the facing page. Third, rasters can be combined with other layers for consistent and superior map display of a surface.

Drawing surfaces in three dimensions

With ArcScene and ArcGlobe, you can make perspective views of an elevation raster derived from a terrain dataset. The elevation raster can be drawn with a color ramp like ███ ██████ , which you can find with other hypsometric color ramps at mappingcenter.esri.com.

For perspective views, you can drape other layers, such as aerial photography and scanned maps, on top of the surface. Using this technique, you can make photo-realistic views of a landscape from any point.

You can combine a series of views along a flight-line into an animation simulating a flight over terrain.

A perspective scene with hillshading and elevation color ramp draped on the surface

Realistic terrains with Swiss hillshading

One way to create a high-quality relief presentation is to download the hillshade toolbar from mappingcenter.esri.com. When you run the Swiss Hillshade tool in the hillshade toolbar, it will launch a geoprocessing model consisting of the datasets and tools shown in this diagram.

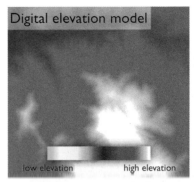

Digital elevation model

low elevation   high elevation

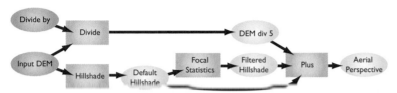

To create a terrain layer, this method produces two rasters from the input digital elevation model (DEM), which is the elevation raster you export from a terrain dataset.

This produces a set of three rasters used to create a Swiss-style hillshading map layer. This implements a technique described in 1965 by Eduard Imhof in his classic book *Cartographic Relief Representation* with modern GIS software.

The input DEM is the digital elevation model, the raster exported from the terrain dataset.

The filtered hillshade is a generalized hillshade that emphasizes the major geographic features, minimizes the minor features, and smooths irregularities on the slopes; however, it still maintains the rugged characteristics of ridge tops and canyon bottoms.

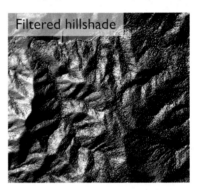

Filtered hillshade

The aerial perspective raster is an additional surface derived from the original DEM and the default hillshade in such a way as to simulate an aerial perspective that makes higher elevations lighter and lower elevations darker.

The three rasters—filtered hillshade, aerial perspective, and DEM—can then be displayed with transparencies to produce an effect similar to the Swiss-style hillshade. Display them in this order: 1) DEM on the top with a color ramp that has variations in hues to show elevation ranges using about 55% transparency, 2) filtered hillshade in the middle with a single hue color ramp, like a grayscale ramp, using about 35% transparency, and 3) aerial perspective on the bottom with a grayscale ramp and no transparency.

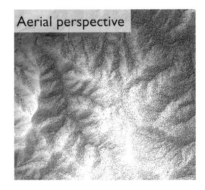

Aerial perspective

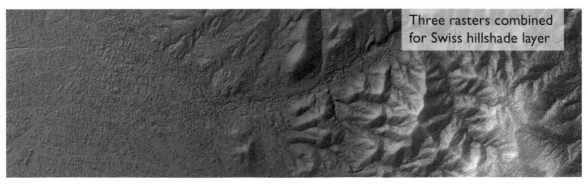

Three rasters combined for Swiss hillshade layer

Building, mapping, and analyzing surface models involves a set of geoprocessing operations. These operations can be summarized in these steps: data collection, feature input, surface models, and analytic rasters.

From features, surface models, and analytic rasters, map layers are made showing a realistic scene using hillshading or surface properties such as slope and aspect. Multiple layers representing terrains can be drawn on a map with the use of transparency. For example, color ramps for elevation rasters are frequently drawn with a transparency on top of a hillshade layer.

Collecting spatial data

This is how photogrammetry works: First, a ground survey is done and around each surveyed point, large white panels forming an 'x' set are set on the ground. These points form a survey network. Next, aircraft with a bottom-mounted camera using large format 9-by-9-inch film capture scenes in succession, timed so that every point in the project area is photographed in two scenes or more, so that a photographic stereo model can be built. These large negatives are mounted on an optical device called a stereoplotter. The photogrammetrist carefully adjusts the orientation of the two plates and then uses a list of map layers and capture rules to digitize surface features in 3D such as contours, building pads, streams, and spot elevations.

Photogrammetry has been used for decades to build surface models and has the advantage of precise definition of features, such as ridge lines and building pads. The disadvantage of photogrammetry is that point collection is limited by the compilation speed and expense of a human operator.

Lidar is replacing photogrammetry for spatial data collection because a lidar survey can generate many millions of elevation points, allowing very fine point spacing for terrain datasets.

Converting data to features

Surface modeling involves the import of large amounts of x, y, z data into features, multipoints catching the bulk of data because of this data type's storage efficiency. Breaklines are used for creases in the landscape such as road edges and streams. Polygons are used for clipping, replacing, erasing, or filling areas in a terrain.

Surface features versus raster sampling

With the input feature classes, you will create a build a terrain dataset, the most accurate representation of the surface possible from the surface data. Terrain datasets have a rich set of analysis tools and cartographic layer properties, but there are many other tools and map layer possibilities with surface rasters.

From terrain datasets, you will commonly export an elevation raster. Doing so provides access to an additional suite of analytic operations and also some improvements for cartographic output, such as surface smoothing.

Surface properties with analytic rasters

From the elevation raster, other rasters can be produced for surface properties of slope, aspect, and curvature. You can analyze a viewshed from a point and make a hillshade raster for realistic surface map layers.

Cartographic output of surfaces

With the terrain dataset and raster surface models derived from it, you have several ways to draw realistic 2D scenes or perspective 3D views.

Use terrain layers for quick display of terrain datasets and when accuracy is paramount, such as engineering applications. Export the terrain dataset to an elevation raster for certain cartographic improvements and additional analysis options.

Surface modeling workflow

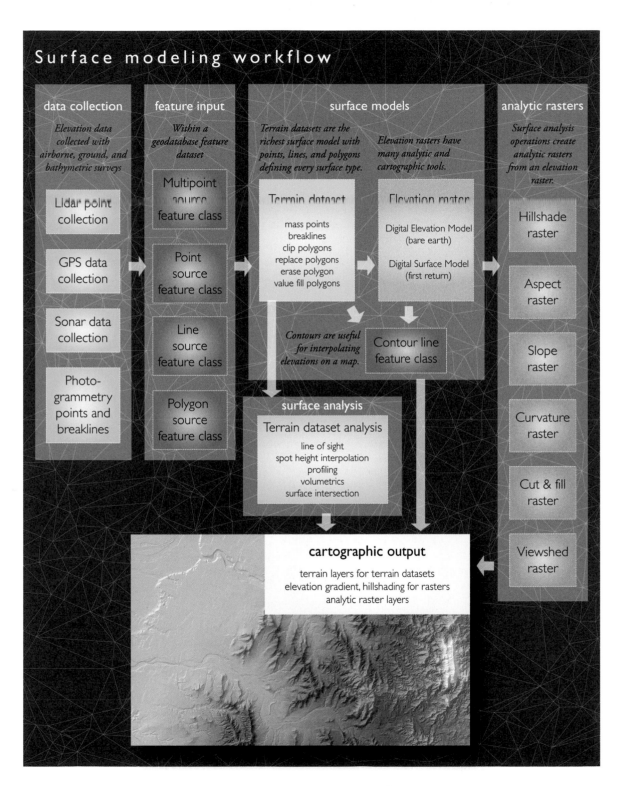

data collection

Elevation data collected with airborne, ground, and bathymetric surveys

Lidar point collection

GPS data collection

Sonar data collection

Photogrammetry points and breaklines

feature input

Within a geodatabase feature dataset

Multipoint source feature class

Point source feature class

Line source feature class

Polygon source feature class

surface models

Terrain datasets are the richest surface model with points, lines, and polygons defining every surface type.

Elevation rasters have many analytic and cartographic tools.

Terrain dataset

mass points
breaklines
clip polygons
replace polygons
erase polygon
value fill polygons

Elevation raster

Digital Elevation Model (bare earth)

Digital Surface Model (first return)

Contours are useful for interpolating elevations on a map.

Contour line feature class

surface analysis

Terrain dataset analysis

line of sight
spot height interpolation
profiling
volumetrics
surface intersection

analytic rasters

Surface analysis operations create analytic rasters from an elevation raster.

Hillshade raster

Aspect raster

Slope raster

Curvature raster

Cut & fill raster

Viewshed raster

cartographic output

terrain layers for terrain datasets
elevation gradient, hillshading for rasters
analytic raster layers

Interactive tools provide the ability to dynamically explore the terrain surface. Geoprocessing tools enable batch like functionality to perform sophisticated geographic analysis.

Conversion tools provide the ability to extract any portion of a terrain surface as a raster. You then use the extracted data with any of the many tools available for rasters.

Sometimes you need to export elevation points to other applications and they cannot handle the large amounts of points in lidar data. When you convert a terrain to points, you can do so at an appropriate pyramid-level for the desired point thinning.

It is possible to bypass the creation of a terrain dataset and create an elevation raster. Point to raster conversion can handle rasterization of multipoints using lidar attributes to create a surface raster with either first return (the Digital Surface Model, DSM) or bare earth (the Digital Elevation Model, DEM).

A variety of analytic operations can be performed on terrain datasets and surface rasters. Some of these operations, such as creating the steepest path, can be done interactively by selecting positions or features on the map. Other operations are done as geoprocessing operations, which can be stored in a geoprocessing model.

The types of surface analysis that operate on terrain datasets (and often on surface rasters) include:

- Interpolate shape analysis calculates z-values based on the terrain surface. Points, lines, and polygons are draped on a terrain dataset or surface raster and have z-values calculated for points and vertices.

- Line-of-sight analysis provides visibility areas. Applications include telecommunications and the protection of scenic views.

- Surface length analysis calculates line length information based on a surface, factoring in z-values along the line.

- Surface spot analysis calculates point height information based on a surface.

- Surface volume analysis calculates the volume of a surface above or below a plane.

- Surface difference analysis subtracts one terrain dataset from another terrain.

- Contours are lines of equal elevation in a landscape.

You can add surface information with a tool that adds surface-related statistics to features. The statistical values made are height for points, surface lengths for 3D lines, minimum and maximum heights for lines and minimum, maximum, and average slope for lines and areas inside polygons.

From terrain datasets and elevation rasters, you can generate analytic rasters that show these surface properties:

- Aspect is the horizontal angle of the surface at a point. Aspect ranges from 0° to 360°.

- Slope is the vertical angle of the surface, measured from the zenith to the normal line projecting from the surface.

- Curvature is the degree the surface varies. It is the second derivative of the surface (slope is the first).

- Cut/Fill analysis compares two surfaces and shows where volumes of material have been moved.

- Hillshading is a technique that uses map aspect and slope for enhancing the depth of 2D maps.

- Observer points analysis finds which areas in a surface are visible to a set of structures, such as towers.

- Viewsheds are areas on a raster visible from a point or along a set of lines.

Creating contours

You can make contours from terrain datasets at any interval. Use contours from terrain datasets for engineering studies and contours from rasters for smoother contours.

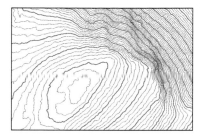

Contours generated from terrain datasets show the fine surface detail.

Finding steepest paths

With points selected on a surface, steepest paths down a terrain can be found. This is the skier's "fall line."

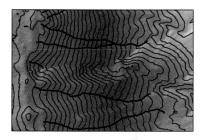

You can see that steepest paths are neatly perpendicular to contour lines.

Line-of-sight analysis

From an observation point, you can derive visibility areas.

The observation point is the white dot. Green areas are visible, while red areas are not.

Interpolating features

Features such as boundaries and roads can be interpolated onto a surface and have z-values calculated.

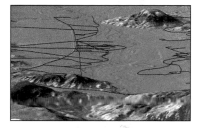

Creating 3D features from 2D geometry

Profile graphs along a line

Along a line, you can make a profile graph showing the elevation changes along that line.

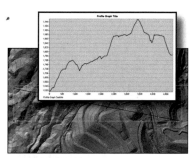

A profile is a graph with the y-axis representing elevation and the x-axis representing distance along the profile line.

Slope of a surface

A slope surface can be calculated from a terrain dataset.

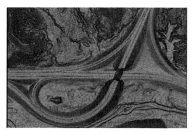

This slope map shows local steepness of terrain. Deep red shows the steepest slopes.

Aspect of a surface

Aspect is the horizontal orientation of a face in the terrain.

Blue shows an aspect to the northwest and red to the south.

Slope and aspect

This aspect-slope map shows the horizontal orientation of the surface (aspect) together with the vertical angle of the surface (slope).

As the color wheel legend shows, aspect is symbolized by color and slope is symbolized by graduated saturation levels of colors.

Surface difference

Elevation rasters can be subtracted to analyze the difference. Examples are reservoir inundation and construction projects.

The elevations of the ground below a reservoir and the volume of water when full

When acquiring lidar data for building a terrain dataset, it's important to consider many environmental variables. This terrain model of a land-water interface was captured by lidar sensors during low tide.

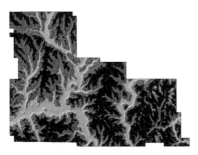

Project area in Broome County, New York, with 1.35 billion mass points from a lidar survey.

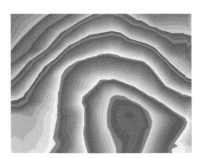

Contour bands rapidly displayed in Dewberry's GeoTerrain application.

There are many applications of terrain datasets in the geodatabase. Urban planners build slope maps to determine which land is useful for residential development or other purposes. Telecommunication companies perform line-of-sight analysis on a landscape to optimize the location of microwave towers. Biologists consider the terrain along with factors such as soil type and vegetation to study wildlife habitats.

One critical application of terrain datasets is the analysis and mapping of floodplain zones. This is an important activity for ensuring public safety and protecting investments within the built environment such as residential housing, public and private facilities, and other critical infrastructure.

Local governments use floodplain maps to determine whether the construction of a home or commercial structure is allowed by statutory regulations. Floodplain maps also affect the availability and cost of property insurance.

Benefits of terrain datasets for floodplain mapping

Dewberry, a multidisciplinary professional services company in the United States, uses terrain datasets to perform floodplain analysis for the Federal Emergency Management Agency (FEMA) and other clients.

Terrain datasets support very large collections of elevation point data. Lidar sensors mounted on aircraft now routinely collect billions of elevation points for an area of interest. TINs work well up to about 10 million points, but are impractical for the massive volumes of elevation points now being collected with lidar sensors. Terrains use multipoint features to efficiently represent many lidar points in a database row and the internal spatial indexing of mass points, so terrain datasets scale well for large amounts of data.

All of the data for a project can be stored in a single repository. Because terrain datasets can efficiently handle large sets of elevation points, there is no need to break a project area into a set of discrete projects, as was common in the past.

Terrain datasets employ pyramid levels, which are progressively thinned sets of elevation points for a series of map scales, so they support seamless and rapid views of surface data.

Modeling floodplains with terrain datasets

Dewberry recognized the utility of terrain datasets for floodplain mapping and built an ArcGIS application using ArcObjects called GeoTerrain.

GeoTerrain can render a surface model as contour bands or lines very quickly on the computer display. Contour generation from large datasets can be time-consuming, but terrain datasets allow the rapid display of contour bands or lines. An analyst can define ranges of elevations to display on-the-fly by manipulating a user interface with terrain symbology. The tool functionality gives the analyst the time and ability to run multiple scenarios that analyze situations such as "What would be the extent of a flooding with a water surface elevation of 124 feet?".

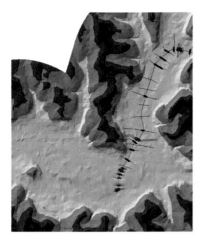

Bridge structures and river cross section lines on a terrain dataset

Profile view of a riverbed and bridge piers for hydraulic analysis

Perspective view of a floodplain displayed on top of aerial imagery

Dewberry performs floodplain analysis for many counties throughout the United States at a number of regional engineering offices. Since these maps and analyses need to conform to national FEMA specifications, Dewberry studied and determined optimal settings for values such as point spacing and pyramid levels and developed a protocol for standardized production of floodplain surface models.

The GeoTerrain application supports visual tools to allow an analyst to view different data source results on the computer display. For example, lidar points can be seen in the same display as field survey points. In this way, ground and bridge data points can be checked between the two data sources before any costly engineering or modeling work has begun.

Before terrain datasets, the limitations of TINs and DEMs (rasters with elevation values) forced a project to be subdivided into tiled areas. A terrain dataset can now model a large project as a single entity and special areas within the project can be analyzed in high resolution, such as fine-tuning hydraulic analysis in problematic areas.

Once the terrain dataset has been built for floodplain analysis, it can be used for other analytic studies such as hydrology, hydraulics, floodplain mapping, risk analysis, and comparative analysis.

Lessons learned

Dewberry has successfully built terrain datasets for many floodplain studies throughout the United States. These are some lessons learned:

- Determine point spacing judiciously. Terrain datasets internally organize mass points into tiles. Each tile may have up to a few million points and their size depends on the point spacing you specify. Be careful to set the point spacing to the average density of points for the area of analysis, which are riverbeds for the GeoTerrain application.

- Use as few pyramids as necessary. For large terrain datasets, each pyramid level takes time to build. Choose only the pyramid level you really need. Most projects at Dewberry use three or four pyramid levels.

- Use z-tolerance pyramids with vertical tolerances of 0.25 to 0.5 feet so that polygons don't have too many unneeded vertices. Floodplain mapping relies on performing a TIN to TIN intersection to create the flood boundary. Each TIN edge that crosses the boundary will create a vertex in the resulting boundary polygon. Not every data point that is important in the vertical direction is important in the horizontal direction. With dense lidar projects, Dewberry finds that 80–90% of vertical data can be filtered from the resultant horizontal polygon without misrepresenting the shape of the floodplain boundary.

Temporal modeling with time-enabled layers

Much of the information in a geodatabase has a time component. There are several forms of temporal data: features that move such as vehicles, features with shapes that change over time such as floodplains, features with attributes that change over time such as land parcels, and rasters that show the state of measurements such as rainfall at snapshots in time. If a dataset contains temporal data, such as a field with a time value, then you can make a time-enabled layer in a map. You can dynamically visualize changes over time in a map with a time- slider. A time-slider in ArcMap lets you visualize your temporal data at selected points in time or with an animated view.

ArcGIS maps can display the changes that occur over time to geographic systems. To enable your maps to show changes in time, you can add temporal data to most geographic datasets, such as feature classes, tables, and rasters.

There are many applications of temporal data. You can analyze weather patterns, changes to ecosystems, wildlife migration patterns, urban development, flooding after major storms, demographic changes, and so on.

Changes over time can be studied over many time extents and intervals. Some long-term changes are studied over decades or centuries, such as how urban development has changed the natural environment. Some temporal data is studied on finer time scales, such as weather data every hour.

Storing temporal data

There are many ways to model temporal relationships in geographic systems. This chapter discusses these general situations with recommendations for adding temporal data to your geodatabase:

- Features that move, such as aircraft and wildlife

- Features at a fixed location with changing attributes, such as stream gauges

- Features that have boundaries that change over time, such as city boundaries and forest fires

- Surfaces and imagery over time, such as rainfall and cloud-cover images

- Attribute data in tables, such as census data values for every decade

You can add time values to most datasets in the geodatabase. For features, time is represented with time values in a feature class. Time values are stored in date, string, or numeric fields with one of several date-time formats. Rasters in a mosaic dataset or raster catalog can also represent a time value with a date field or string field.

Visualizing temporal data

When you add a layer to a map, you can make that layer time-enabled by setting time properties in the Time tab in the Layer properties dialog box. You can have one or many time-enabled layers on a map.

For each layer, you specify either one timestamp field or, if modeling a time duration, a begin and end time field. A layer with time properties also has a temporal reference, which specifies the time zone for time values. The temporal reference lets you combine layers from different time zones.

Once you have time-enabled layers in a map, you can visualize your temporal data through a time-slider. A time-slider is a control on the map which lets you view your temporal data for any point in time or as an animated view. You can control the time extent (the begin and end time to view your temporal data) and the time interval (such as hours, days, and months) to display your temporal data.

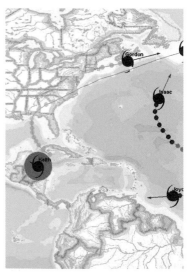

Analyzing hurricane tracks in the Atlantic Ocean

The maps in this topic show several applications of temporal data as presented by ArcGIS users at the annual ESRI User Conference.

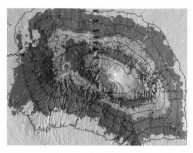

Extent of forest on Mt. Kilimanjaro, Kenya
1976

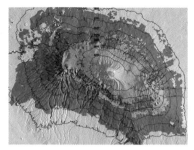

Extent of forest on Mt. Kilimanjaro, Kenya
2000

The reduction of rainfall combined with increasing temperatures has caused changes in the ecosystems of Mt. Kilimanjaro. Not only is the ice cap disappearing, but the forest boundary is shifting significantly and becoming susceptible to fires.

This map shows the changes to ecosystems over time and illustrates the effects of climate change.

The Jacksonville Police Department wanted to study juvenile crime patterns over space and time. This map shows arrests of people younger than 18 during school days. Displaying crimes by time range showed patterns in criminal activity in the vicinity of schools, prompting the police to step up targeted patrols.

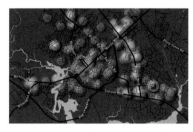

After-school crime density from 2:16 PM to
4:00 PM

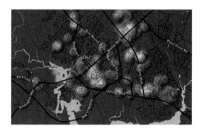

Evening crime density from 4:00 PM to
7 AM

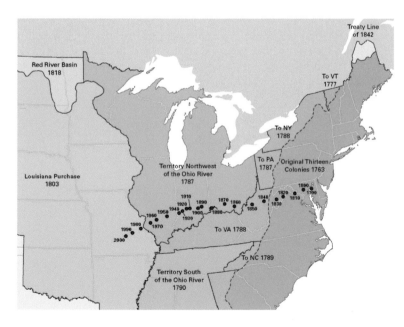

Once every decade, the U.S. Census Bureau performs an enumeration of the population. As part of the tabulation and publication activities after each census, the country's center of population is calculated. This center is calculated to be the point of balance of a map of the United States (using boundaries at the time of each census) that gives equal weight to all residents at their geographic locations.

This map shows the expanding territory of the United States and the steady progression of the center of population from the first census of the original colonies to the census of 2000. The center of population's shift towards the west and south reflects both the expansion of the national territory and migration patterns towards western and southern states.

Some features move over time, such as vehicles on highways, aircraft on air corridors, and animals on migration patterns. You can model moving features from data in feature classes or in real time using network connections to GPS receivers.

The maps to the right show air traffic data and how it is displayed with a time-slider. In this case, air traffic data is stored in a point feature class, with an extract of the attribute values shown below. In this feature class, the positions of aircraft are collected once a minute and the time is recorded in the field called TimeMessage. When adding this layer to a map, you can make it time-enabled by identifying TimeMessage as the time field.

A flight follows a track and a successive set of points can define this track. In this case, the CallSign attribute defines the track. When you define a track in a time-enabled layer, you have options for displaying the related points, such as drawing directional vectors or displaying recent positions, as shown on the maps to the right. In these maps, the symbols for current aircraft position was oriented by using the Heading attribute, which stores the aircraft bearing in degrees relative to the north direction.

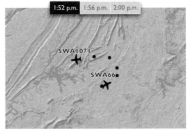

This is a view of the flight data at a timestamp after flights SWA1071 and SWA66 have departed. Symbols for aircraft are drawn at the current timestamp in the time-slider.

AirTraffic

Shape	CallSign	Origin	Destination	Altitude	Speed	Heading	TimeMessage
Point Z	SWA1071	ATL	MEM	027	235	060	4/4/2008 1:51:00 PM
Point Z	SWA1071	ATL	MEM	044	348	065	4/4/2008 1:52:00 PM
Point Z	SWA1071	ATL	MEM	077	250	007	4/4/2008 1:53:00 PM
Point Z	SWA1071	ATL	MEM	099	270	321	4/4/2008 1:54:00 PM
Point Z	SWA1071	ATL	MEM	111	236	273	4/4/2008 1:55:00 PM
Point Z	SWA66	ATL	OKC	020	240	060	4/4/2008 1:54:00 PM
Point Z	SWA66	ATL	OKC	048	269	075	4/4/2008 1:56:00 PM
Point Z	SWA66	ATL	OKC	071	275	358	4/4/2008 1:57:00 PM
Point Z	SWA66	ATL	OKC	097	301	353	4/4/2008 1:58:00 PM
Point Z	SWA66	ATL	OKC	120	309	349	4/4/2008 1:59:00 PM
Point Z	SWA1097	ATL	OAJ	030	224	060	4/4/2008 1:56:00 PM
Point Z	SWA1097	ATL	OAJ	050	236	090	4/4/2008 1:57:00 PM
Point Z	SWA1097	ATL	OAJ	072	252	095	4/4/2008 1:58:00 PM
Point Z	SWA1097	ATL	OAJ	093	277	097	4/4/2008 1:59:00 PM
Point Z	SWA1097	ATL	OAJ	113	311	101	4/4/2008 2:00:00 PM

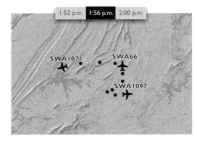

This is a view at a timestamp several minutes later. Flight SWA1097 has joined the other two flights.

You can also represent and map features that move or change in real-time. With the ArcGIS Tracking Analyst extension, you can establish network connections to Global Positioning System (GPS) units and other tracking and monitoring devices so you can map your data in real time. Tracking Analyst accepts real-time data in three forms:

- Simple events, which have a position and time with other attributes.

- Complex dynamic events, which model a moving sensor location with incoming temporal observations. Aircraft observations are an example.

- Complex stationary events, which model a fixed sensor location with incoming temporal observations containing time and status attributes. An example is monitoring a traffic sensor.

Read the ArcGIS help system for more information on the Tracking Analyst extension.

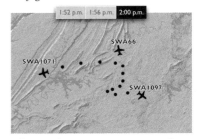

After several more minutes, you can see the three flights diverging to their destinations.

Some features change shape over time. The features shown on these maps are inundation levels of a reservoir over time.

Since the shape of the reservoir level changes at every timestamp value, a new feature is created in the feature class for each timestamp. A feature that changes shape can be specified by using a common identifier or name. In this example, one geographic object, the Chattahoochee Reservoir, is tied to five features with changing shape by the Name field.

In the feature class for reservoir levels, the inundation level is modeled once every several months. The feature for that level is given a timestamp with year and month designation. It is not necessary to specify a fine-grained time value such as day, hour, minute, and seconds if that is not appropriate for your application. Timestamps also need not occur at a regular time interval. They can represent any arbitrary point in time of interest to you.

ReservoirLevel

Shape	Name	Level	TimeStamp
Polygon	Chattahoochee Reservoir	320	2006/03
Polygon	Chattahoochee Reservoir	312	2006/06
Polygon	Chattahoochee Reservoir	324	2006/10
Polygon	Chattahoochee Reservoir	317	2006/12
Polygon	Chattahoochee Reservoir	323	2007/04

Two other common applications of features that change shape are flood limits during a major rainfall event and fire boundaries. These dynamic events will be represented with a finer time granularity, perhaps once an hour. For fire maps, graduated colors can be used to clearly show the progression of a fire over time.

This perspective view of a fire in the San Bernardino mountains shows the perimeter on November 1, 2003. The perimeter was captured using aerial equipment and GPS receivers and overlaid on a digital elevation model to produce this perspective.

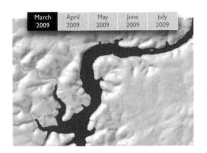

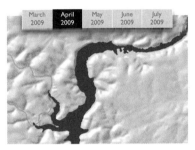

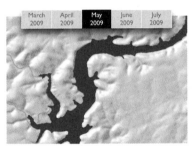

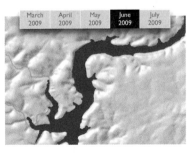

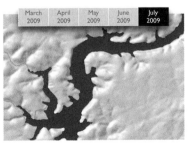

An example of modeling features with attributes that change over time are stream gauges which record flood levels. These maps were made by adding a stream gauge feature class layer to a map and enabling the temporal display by setting time properties for the layer.

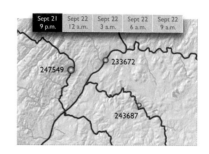

One possible approach is to model multiple features for each moment in time, as shown in the table below. If five timestamps are collected for three stream gauges, then your point feature class could have 15 point features for each combination of stream gauge and timestamp.

The feature class design below can work if your feature class is generated from other data, but this design introduces data redundancy and can be difficult to maintain.

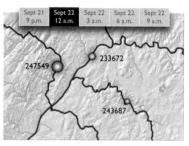

▫▪ StreamFlow

Shape	GaugeID	GaugeName	GaugeHeight	TimeStamp
Point	233672	Chattahoochee River near Fairburn	2.58	9/21/2009 9:00:00 PM
Point	233672	Chattahoochee River near Fairburn	4.53	9/21/2009 12:00:00 AM
Point	233672	Chattahoochee River near Fairburn	5.21	9/21/2009 2:00:00 AM
Point	233672	Chattahoochee River near Fairburn	12.63	9/21/2009 6:00:00 AM
Point	233672	Chattahoochee River near Fairburn	17.21	9/21/2009 9:00:00 AM
Point	243687	Sweetwater Creek near Austell	1.20	9/21/2009 9:00:00 PM
Point	243687	Sweetwater Creek near Austell	2.36	9/21/2009 12:00:00 AM
Point	243687	Sweetwater Creek near Austell	4.58	9/21/2009 2:00:00 AM
Point	243687	Sweetwater Creek near Austell	6.33	9/21/2009 6:00:00 AM
Point	243687	Sweetwater Creek near Austell	5.39	9/21/2009 9:00:00 AM
Point	247549	Nickajack Creek at US 78/278	4.56	9/21/2009 9:00:00 PM
Point	247549	Nickajack Creek at US 78/278	8.55	9/21/2009 12:00:00 AM
Point	247549	Nickajack Creek at US 78/278	10.98	9/21/2009 2:00:00 AM
Point	247549	Nickajack Creek at US 78/278	19.43	9/21/2009 6:00:00 AM
Point	247549	Nickajack Creek at US 78/278	21.02	9/21/2009 9:00:00 AM

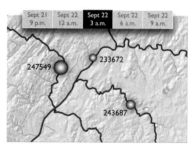

A better approach is to represent stream gauge data with a one-to-many relationship between features in a point feature class and rows in a table. To do temporal display with this feature layer, you would add a join from the feature class with stream gauges to a table with gauge height measurements. In this example, the GaugeID is the identifier that links the two tables.

▤ GaugeHeight

GaugeID	GaugeHeight	TimeStamp
233672	2.58	9/21/2009 9:00:00 PM
233672	4.53	9/21/2009 12:00:00 AM
233672	5.21	9/21/2009 2:00:00 AM
233672	12.63	9/21/2009 6:00:00 AM
233672	17.21	9/21/2009 9:00:00 AM
243687	1.20	9/21/2009 9:00:00 PM
243687	2.36	9/21/2009 12:00:00 AM
243687	4.58	9/21/2009 2:00:00 AM
243687	6.33	9/21/2009 6:00:00 AM
243687	5.39	9/21/2009 9:00:00 AM
247549	4.56	9/21/2009 9:00:00 PM
247549	8.55	9/21/2009 12:00:00 AM
247549	10.98	9/21/2009 2:00:00 AM
247549	19.43	9/21/2009 6:00:00 AM
247549	21.02	9/21/2009 9:00:00 AM

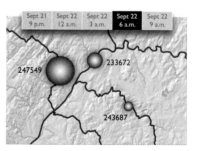

▫▪ StreamGauge

Shape	GaugeID	GaugeName
Point	233672	Chattahoochee River near Fairburn
Point	243672	Sweetwater Creek near Austell
Point	247549	Nickajack Creek at US 78/278

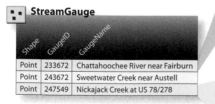

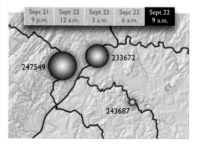

Rainfall totals as measured from satellite sensors is an example of modeling surfaces that change over time. A similar application is displaying satellite imagery showing cloud cover on a daily basis.

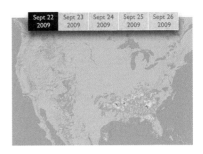

These maps show daily rainfall totals for 24-hour periods. Each day's rainfall totals are represented by rasters in a mosaic dataset or raster catalog. The only requirement to enable the time-visualization of this data is to add time fields to the table for your mosaic dataset or raster catalog.

In this example, a raster catalog has two fields with time values: Start_Time and End_Time. These two fields specify the time intervals for which this data is valid. When this raster catalog is displayed using a time-slider, a raster will remain visible for every time instant between the start and end time values.

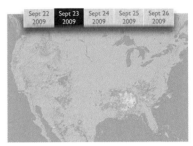

⊞ Precipitation

Raster	Name	Start_Time	End_Time
<Raster>	tprecip0	9/20/2009 12:00:00 AM	9/22/2009 12:00:00 AM
<Raster>	tprecip1	9/21/2009 12:00:00 AM	9/23/2009 12:00:00 AM
<Raster>	tprecip2	9/22/2009 12:00:00 AM	9/24/2009 12:00:00 AM
<Raster>	tprecip3	9/23/2009 12:00:00 AM	9/25/2009 12:00:00 AM
<Raster>	tprecip4	9/24/2009 12:00:00 AM	9/26/2009 12:00:00 AM

When you add a layer to a map with a mosaic dataset or raster catalog, you can enable the temporal display of this data with a time-slider by setting the time properties of the layer.

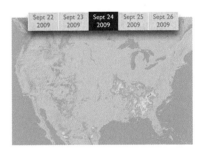

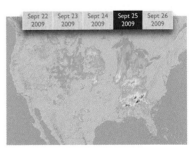

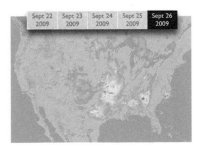

NetCDF (Network Common Data Form) is a data format and software library designed to store and access multidimensional arrays of environmental data. NetCDF was developed by the University Corporation for Atmospheric Research (UCAR) and is widely used for atmospheric modeling and oceanography.

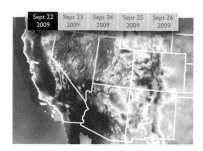

The netCDF Web site, http://www.unidata.ucar.edu/software/netcdf/, describes these key attributes:

NetCDF data is:

- Self-Describing. A netCDF file includes information about the data it contains.

- Portable. A netCDF file can be accessed by computers with different ways of storing integers, characters, and floating-point numbers.

- Scalable. A small subset of a large dataset may be accessed efficiently.

- Appendable. Data may be appended to a properly structured netCDF file without copying the dataset or redefining its structure.

- Sharable. One writer and multiple readers may simultaneously access the same netCDF file.

- Archivable. Access to all earlier forms of netCDF data will be supported by current and future versions of the software.

Who uses NetCDF?

NetCDF is a standard format for data distribution to share earth environmental data. Among some of the U.S. agencies that produce data in NetCDF form are NOAA (National Oceanic and Atmospheric Administration), NASA (National Aeronautic and Space Agency), the U.S. Navy, NCAR (National Center for Atmospheric Research), and NSIDC (National Snow and Ice Data Center). NetCDF is also widely used by environmental agencies of governments around the world, such as Australia, France, Italy, South Korea, and the Netherlands.

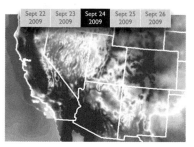

If your GIS application requires the analysis of the earth's atmosphere and oceans, there is a good chance that you'll find this data in netCDF format.

NetCDF and ArcGIS

In ArcGIS, you can create a layer in ArcMap which references a netCDF file. NetCDF files can contain raster, feature, or tabular data. NetCDF files are called multidimensional because they usually reference these dimensional values: latitude, longitude, height, and time. By its design, netCDF can handle any number of environmental attributes, such as rainfall, barometric pressure, windspeed, and others. You can select which of these attributes are displayed in your netCDF layer.

Since netCDF data has a time-component, netCDF layers can be made time-enabled and visualized with the time-slider and animations.

The maps on the right show netCDF layers of daily temperatures.

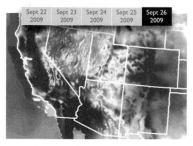

County

Shape	Name	Pop1980	Pop1990	Pop2000
Polygon	Rio Arriba	28,767	30,430	32,411
Polygon	Taos	35,940	37,874	39,566

This feature class contains counties with population values by decade. Because it has multiple fields with values for different times, it cannot be used for temporal visualization with a time-slider.

County

Shape	Name	Year	Population
Polygon	Rio Arriba	1980	28,767
Polygon	Rio Arriba	1990	30,430
Polygon	Rio Arriba	2000	32,411
Polygon	Taos	1980	35,940
Polygon	Taos	1990	37,874
Polygon	Taos	2000	39,566

This feature class contains the same data, but reformatted so that only one field represents a temporal value. This is the recommended form for temporal visualization.

To eliminate data redundancy, you can also separate this data into a feature class with a join to a table containing the population statistics. But if there were any changes in county boundaries over decades, then the single table form as shown above is best.

Temporal data can be stored in your geodatabase in a variety of ways. To visualize your temporal data on a map with a time-slider, these are the recommended best practices.

Store temporal data in a row format

You should store temporal data in a row format, which means that each feature or raster has either one field for the timestamp (date and time combined), or two fields for the begin and end time values.

Sometimes you'll encounter data that stores temporal values in multiple fields. An example would be a feature class representing administrative areas with population values for successive decades, such as 1980, 1990, 2000, and so on. An example is shown in the feature classes to the left.

To visualize this data with a time-slider, it will be necessary to reformat the table such that each row represents a population value for one decade. ArcGIS contains a software tool to transpose time fields to reformat tables with multiple time fields.

Store time values in a Date field

You can store time values with either a date field, string field, or numeric field, but date fields are recommended. Date fields allow the best query performance and supports more sophisticated database queries.

Index fields that contain time values

For the best performance in time visualization and query, you should add attribute indexes for the fields that contain time values. Attribute indexes are used by ArcGIS to quickly locate records that match an attribute query.

Use standard time

You should store temporal values using standard time in a time zone and not use local daylight savings time.

Many countries around the world use daylight savings time during the summer months. This is the practice of shifting the clock by one hour to maximize daylight during waking hours. While the time-slider can display temporal values using daylight savings time, you should not store temporal data using daylight savings time because of the potential for confusion, especially during the times when local time switches from standard time to daylight savings time and back. Most agencies that record temporal data do not use daylight savings time in their time values.

You can visualize temporal data in ArcGIS by setting time properties on one or more layers in a map. Once you've added a dataset with time information as a layer to a map, you can make it a time-enabled layer. With time properties set, you can use a time-slider in ArcMap to examine and visualize how the features or rasters in that layer change over time.

Time properties of time-enabled layers

Each piece of data (feature, raster, or other) in a time-enabled layer has a timestamp. A timestamp can either represent an instant in time or a time interval between two instants in time.

July 20   July 21   July 22

time instant   *time interval*

A timestamp can be stored in several field types: date, string, or numeric. It is recommended to use date fields for best display and analysis performance. You can use string and numeric fields if you use time values created from information systems outside ArcGIS.

Sometimes, tables will have a time value divided into two or more fields: one field for years, months, and days and another field for hours, minutes, and seconds. Before you work with this temporal data in a time-enabled layer, you must reformat these separate time fields into one time field.

A time-enabled layer can have either a single time field or a start and end time field. Use a single time field if you want that data to only be displayed at that time instant. Use start and end time fields if you want that data to be displayed over an interval of time.

A time-enabled layer has a time zone. You can select any of the major world time zones. Specifying a time zone as a property of a time-enabled layer enables the time-slider to integrate multiple time-enabled layers with different time zones. A time-slider also has a time zone and converts and displays time data to this time zone.

A time-enabled layer has a time interval and a time extent. A time interval is the period of time between two adjacent time values, typically, a year, month, day, hour, or minute. A time extent is the period of time between the first timestamp and last timestamp for the time-enabled layer.

Time properties of the time-slider

The time-slider provides controls to visualize temporal data in ArcGIS. The following properties can be set on a time-slider:

- You can set the time step interval of the time-slider, such as hours, days, or months.

- You can set the time extent of the timeline. It can match the time extent of the time-enabled layers or any time extent you specify.

- You can set the time zone in which you want to visualize your data. All time-enabled layers will have time values translated to this time zone.

- You can specify the format that you want to display timestamps, such as 'September 9, 2009'.

- You can set play options for the time-slider, such as duration in seconds. Other options are to reverse or repeat the display.

The illustration on the facing page shows the integration of three time-enabled layers and how they are visualized with the time-slider.

Supported string field formats
YYYY
YYYYMM
YYYY/MM
YYYY-MM
YYYYMMDD
YYYY/MM/DD
YYYY-MM-DD
YYYYMMDDhhmmss
YYYY/MM/DD hh:mm:ss
YYYY-MM-DD hhmmss
YYYYMMDDhhmmss.s
YYYY/MM/DD hh:mm:ss.s
YYYY-MM-DD hh:mm:ss.s

Supported numeric field formats
YYYY
YYYYMM
YYYYMMDD
YYYYMMDDhhmmss

While date fields are best for storing time data, sometimes you might need to use string or numeric fields for compatibility with data generated by other systems. These are the string and numeric field formats that can be used in defining time-enabled layers.

Streamflow time-enabled layer

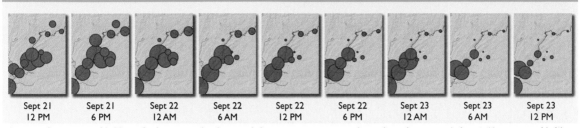

| Sept 21 12 PM | Sept 21 6 PM | Sept 22 12 AM | Sept 22 6 AM | Sept 22 12 PM | Sept 22 6 PM | Sept 23 12 AM | Sept 23 6 AM | Sept 23 12 PM |

The streamflow time-enabled layer displays stream heights recorded at a stream gauge every hour, shown here every six hours. This time-enabled layer has a time step interval of six hours and one time field with a value for every hour.

Wind direction time-enabled layer

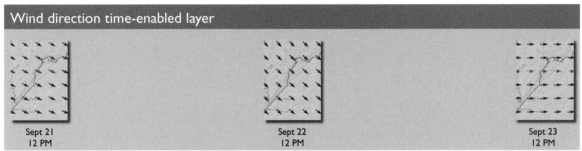

| Sept 21 12 PM | Sept 22 12 PM | Sept 23 12 PM |

The wind direction time-enabled layer displays gridded points with attributes for wind direction and velocity recorded once a day. This time-enabled layer has a time-step interval of one day and one time field with a value for every day at 12 PM.

Precipitation time-enabled layer

| Sept 21 12 PM | Sept 22 12 PM | Sept 23 12 PM |

The precipitation time-enabled layer displays a raster catalog with rasters representing accumulated rainfall for each day. This time-enabled layer has a time-step interval of one day and two time fields. The start time field goes from 12 PM one day and the end time field goes to 12 PM the next day.

Combined streamflow, wind direction, and precipitation time-enabled layers

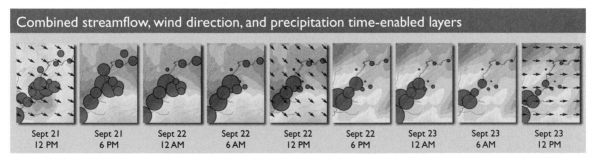

| Sept 21 12 PM | Sept 21 6 PM | Sept 22 12 AM | Sept 22 6 AM | Sept 22 12 PM | Sept 22 6 PM | Sept 23 12 AM | Sept 23 6 AM | Sept 23 12 PM |

The combined time-enabled layers are displayed with a time-slider set to a time-step interval of six hours. The streamflow layer is updated every six hours. Because the wind direction layer only has one time field, instead of a begin- and end-time field, it is only displayed at 12 PM each day. The precipitation layer is displayed every six hours within the one-day interval between the start and end time field values.

Multiuser editing with versioning

Versioning allows multiple users to edit the same data in an ArcSDE geodatabase without applying locks or duplicating data. A version is a private view of a geodatabase that can represent an engineering design, a construction job, or any other type of transaction. Any edits that you do within a version are isolated from any other person's view of the geodatabase until you decide to post your changes. If you need to work with data in separate locations, you can also distribute all or part of the data in a geodatabase through replication. Replicas of geodatabases are used for situations such as remote field offices or field work with a laptop computer.

A key aspect of working with a modern GIS is the ability to manage editing workflows in the geodatabase. The management of editing workflows is especially important for enterprise GIS systems in which many editors are simultaneously working with the same geographic information.

Consider an electric utility company that has multiple editors who concurrently access the same datasets. These editors are adding, removing, and updating features as the utility networks change. The geodatabase accommodates this requirement through versioning by expanding the underlying transaction model of the relational database.

Versioning is a core function of ArcSDE geodatabases because it facilitates these multiuser workflows by supporting a long transaction framework for managing many GIS editing and data maintenance scenarios.

Versioning is also the technology behind geodatabase replication that allows for distributed GIS operations, as well as geodatabase archiving, which maintains a historical record of editing transactions.

How versioning works

A geodatabase version can be thought of as a unique and local view of a geodatabase. Once you've created a version, you have made what appears to be your own private snapshot of the data. This version can then be edited separately from the current default representation of the geodatabase.

Contrary to appearances, you have not copied any data when you establish a version. Rather, a version uses tables associated with versioned datasets, called delta tables, to track any changes made in the version. After you've made edits in a version you can merge them into the default representation of the geodatabase, overwriting the current representation of the data with the tracked changes that you've made.

The illustration on the facing page displays an example of how two GIS editors, Polly and Michael, simultaneously make edits to a parcel feature class. Polly is merging two parcels into one parcel, while Michael is splitting a nearby parcel into four smaller parcels. Each is working within their own versions created from the default representation of the geodatabase, and their edits are only visible to themselves.

After Polly finishes her edits, she posts the changes to the default version. The information in the delta tables, showing the more recent representation of the data, is used to update the default version.

Next, Michael finishes his edits and also posts his changes to the default version. In this example, there are no conflicts between his edits and Polly's edits, so merging the updates goes smoothly. If there were conflicts, such as Polly and Michael making edits to the same parcel, then there is a conflict management process to inspect and reconcile differences between versions. That's discussed later in this chapter.

When to use versioning

If you want to take advantage of the following functionality, then you must use versioning in an ArcSDE geodatabase:

- Managing updates on datasets coming from multiple simultaneous editors.

- Editing features in an ArcSDE geodatabase that participate in data structures such as geometric networks, network datasets, terrain datasets, and topologies.

- Distributed GIS operations through geodatabase replication, including disconnected editing in the field and synchronizing replicas stored across different relational databases.

- Creating and maintaining a historical record of datasets using geodatabase archiving.

An ArcSDE geodatabase always has a top-level version that is always called 'default'. The default version is the current representation or public view of the data in the geodatabase. In the illustration on the facing page, the default version presents the land cadastre as it exists at the current time.

Edits made to a versioned dataset are tracked in two tables called the delta tables. By tracking changes in these tables, the versioning architecture does not have to apply locks to any data while it is being edited, nor does it have to duplicate any data by making a separate copy available for editing.

A geodatabase has 14 land parcels within a subdivision. These are the parcels in the default version of the geodatabase. Polly and Michael work in a county assessor's office and are responsible for updating the land cadastre when real estate transactions take place. They each create their own version from the default version and make edits to parcels in the same subdivision.

Feature class **Parcel**

Parcel ID	Address	Shape
279	24 Lopez Drive	Polygon
281	27 Schweitzer Place	Polygon
283	1314 White Road	Polygon
321	1306 White Road	Polygon
322	22 Schweitzer Place	Polygon
323	24 Schweitzer Place	Polygon
324	32 Lopez Court	Polygon
326	30 Lopez Court	Polygon
381	23 Lopez Drive	Polygon
382	25 Lopez Drive	Polygon
383	27 Lopez Court	Polygon
384	29 Lopez Court	Polygon
385	31 Lopez Court	Polygon

Polly connects to her version and merges parcels 384 and 385 into parcel 386.

Michael connects to his version and splits parcel 283 into four parcels: 284, 285, 286, and 287.

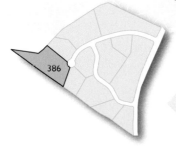

Polly posts her updates back to the default version.

Michael also posts his updates to the default version. In this example, no conflicts are encountered.

The default version of the geodatabase now has 16 land parcels within the subdivision. The default version is brought up-to-date by incorporating recent real estate transactions made from Polly's and Michael's simultaneous edits in their respective versions.

Feature class **Parcel**

Parcel ID	Address	Shape
279	24 Lopez Drive	Polygon
281	27 Schweitzer Place	Polygon
284	1314 White Road	Polygon
285	1312 White Road	Polygon
286	1310 White Road	Polygon
287	1308 White Road	Polygon
321	1306 White Road	Polygon
322	22 Schweitzer Place	Polygon
323	24 Schweitzer Place	Polygon
324	32 Lopez Court	Polygon
326	30 Lopez Court	Polygon
381	23 Lopez Drive	Polygon
382	25 Lopez Drive	Polygon
383	27 Lopez Court	Polygon
386	29 Lopez Court	Polygon

Versioning is only available with ArcSDE geodatabases. You cannot implement versioning on file or personal geodatabases.

Versioning lets multiple users directly edit an ArcSDE geodatabase without applying feature locks or duplicating data.

In versioned editing workflows, each simultaneous editor works with their own individual geodatabase version, a snapshot of all the versioned information in a geodatabase. An ArcSDE geodatabase can have many versions and every geodatabase starts with a version named default that is owned by the ArcSDE administrator.

ArcSDE geodatabase editors add versions to their geodatabase as required for projects. Each editor interacts with their data through a specific version.

You can use versions to represent engineering designs, construction jobs, maintenance activity, updates to land parcels and their ownership, and any collection of changes performed by an editor.

New versions are always created from an existing version. The first version you create in an ArcSDE geodatabase is a child of the default version. The default version is therefore an ancestor to all versions in a geodatabase.

Properties of a version

Every version in a geodatabase has an owner, a description, and a permission level. The owner is the geodatabase user who first created the version. The owner can add a textual description of the version as well as set the specific permissions that the version has. Permissions on a version can be used to control other users' access to that version. Assigning version permissions is one of the key tools used to control specific workflows in a multiuser editing environment.

The three possible permission levels of a version are:

- Private: Only the owner of the version can view and edit datasets within that version.

- Protected: All users can view datasets within that version but only the version owner can edit datasets.

- Public: All users can view and edit datasets within that version.

The delta tables are sometimes called the A (adds) and D (deletes) tables. These two tables work together to track changes made to a versioned dataset in an enterprise geodatabase. Since editors are working within separate versions, multiple editors can edit the same features without locking them to other editors. Also, no data is duplicated since all edits are tracked in the delta tables instead of creating copies of the data for each editor to work with.

A key point to this is that all records in the delta tables reference what specific version the change was made in. ArcGIS uses this information to display the appropriate representation of a versioned dataset to the end user, depending on the version they are connected to. In this way, a version acts as an isolated snapshot of the datasets in a geodatabase.

How changes are tracked with versioning

Versioning works by simply tracking all the changes that are made in a geodatabase. These changes are tracked in a set of tables which are collectively referred to as the delta tables. Every dataset that is registered as versioned has their own set of delta tables. As geographic information is created, updated, and deleted over time, records are added to the delta tables tracking these specific edits. Through the delta tables, versioning provides users with an independent view of edits made to their geographic data.

When working in a version, any geodatabase dataset that you wish to edit must first be registered as versioned. Registering a geodatabase dataset as versioned creates the two delta tables in the geodatabase and associates those tables with the dataset.

The illustration below shows how versioning works. You can see how delta tables get updated when adding, deleting, and modifying a feature.

Once a dataset is registered as versioned any edits that are made to that dataset will be tracked in the delta tables. For example, if a feature is inserted into that feature class a record will be entered in the adds table associated with the feature class. Information such as the feature's attributes and what specific version the edit was made on will be entered into the adds table.

If a feature is deleted in the feature class, instead of inserting a record in the adds table, a record will be inserted into the deletes table. To track a deletion, all the geodatabase needs to know is what version the feature was deleted on.

Finally, when a feature is updated in the feature class the versioning infrastructure will insert a record into the deletes table and also insert a record into the adds table. This will effectively retire the old representation of the feature (and add a row in the deletes table) and replace it with the new representation of that feature (and add a row to the adds table).

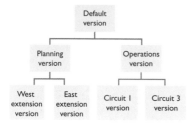

You can design a version tree of any number of levels appropriate for your organization's workflow. For best performance and to simplify administration, it is better to have a flatter version tree than a deeper version tree.

Geodatabase versioning supports a variety of different workflow processes. In most organizations, workflows generally progress in discrete stages, each stage representing an assignment of work or a project. To manage each work assignment, a separate version can be created and edited until the work is completed and the edits are ready to be incorporated into the default version of the geodatabase. By administering projects in this manner, versioning offers a data management framework flexible enough to accommodate many different workflows.

These are examples of common versioning tasks that can be done by organizations while supporting simultaneous edits by multiple GIS users:

- Updating the address of a customer in a utility database

- Subdividing a parcel to reflect a real estate transaction in a parcel database

- Adding a service to a new customer in a utility database

- Updating a forest block to reflect a planned cutting operation

- Designing a new substation in a utility database

- Checking out a section of a utility database, modifying it in the field to reflect storm-related damage, and checking the work back in to the central database

- Planning a new subdivision in a land planning database

- Performing a what-if scenario for a disaster recovery simulation

A version tree is a general term used to describe the relationships between versions in a geodatabase. The structure of a version tree is often best suited to a specific workflow. Three generalized examples of version trees are shown in the following graphics, depicting some common workflows addressed by versioning.

Concurrent editing of the default version

Directly editing the default version is the simplest method of implementing multiuser editing. In this workflow, each editor does not need to explicitly create a new version. When an edit session is started on the default version, the geodatabase creates an unnamed, temporary version to isolate your edits from other changes made to the default version. This temporary version is only accessible to the current editor.

When the editor saves their work or ends the edit session, edits from the temporary version are merged into the default version. If the editor decides not to save their edits the version will be deleted and any changes made in the edit session will be discarded.

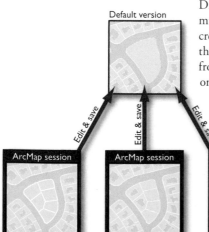

Multiple projects

If you're managing multiple projects or work orders, you'll require a more structured approach to workflow management. As projects and work orders vary in length, changes made to each version can involve many edit sessions spanning a number of days, weeks, or months. Examples of these discrete work units could be a highway improvement project, the installation of a new phone service, or an ongoing maintenance project for a gas pipeline.

When a work order or project is initiated, a version is created as a child of the default version. One or more editors can work on this version until the work order or project is complete. When all modifications to that version have been completed, the editor integrates the changes made on the child version with the default version. At the completion of the project, the version which represented the project can optionally be removed from the geodatabase.

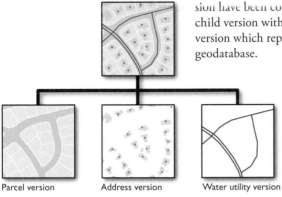

Default version

Parcel version   Address version   Water utility version

With this workflow it may be necessary to restrict the user access permissions on the default version. This will ensure that the default version is not modified by anyone other than the ArcSDE administrator. To do this the ArcSDE administrator may set the permission of the default version to protected, allowing users to continue to view the default version but restricting their access level to read-only. In this case, any editor wishing to modify the data must create a new version.

Multiple projects with a quality control version

A variation on the multiple project workflow is to use a quality control version as an integration stage for other versions. This is another way of protecting the default version from direct editing and unauthorized or unintentional modification.

In this scenario all editors would merge their changes with the quality control version and a quality assurance manager would then have control over the edits that get integrated back to the default version. Since this quality control version is a child of the default version, this is often referred to as a surrogate default version.

First, a new quality control version would be created from the default version. All other versions are then created from the quality control version as required. In this workflow, editors would never interact with the default version directly. This isolates any changes made by the editors from the

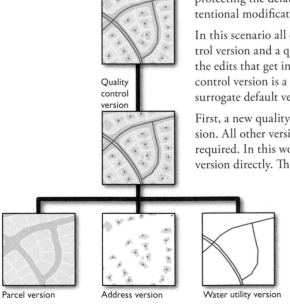

Default version

Quality control version

Parcel version   Address version   Water utility version

default version of the geodatabase until such time as it is considered appropriate to integrate these changes. A database administrator or quality assurance manager can then merge the changes from the quality control version, resolve any conflicts, and integrate the edits into the default version without affecting other database users.

For the most part, editing versioned data in an ArcSDE geodatabase is not much different from editing nonversioned data in a file or personal geodatabase. The user connects to the data in ArcGIS. When connecting to data in an ArcSDE geodatabase, you specify which version you want to connect to.

One unique detail of versioned editing is that the editing workflow also involves the merging of changes between versions in a geodatabase. Editors make changes in isolated versions until they are ready to incorporate those edits into the default version or another version. Versioned editing provides a mechanism for merging the changes made in a child version into any of that version's ancestors. This process is called reconcile and post and is a critical part of the version editing process which makes the workflows discussed in the previous section possible.

Reconciling versions

The first step in merging edits from one version to another is a process called reconcile. When versions are reconciled with one another, edits in the target version are merged with those made in the current version being edited, called the edit version.

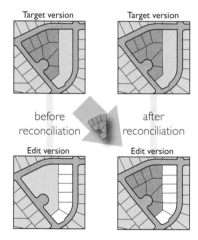

During the reconcile process, edits to the parcel features made in the target version get pushed into the edit version.

The reconcile process will also notify the editor of any conflicts that might arise during the merging process. A conflict arises any time the same feature or row has changed on both the target and edit versions. Since versioning does not apply locks to the data as it is being edited there is nothing stopping different editors from modifying the same feature in their respective versions. Identifying possible conflicts during the reconcile process allows the editor to make a decision as to which representation of the feature is valid. There are two types of conflicts which can arise during the editing process:

1. When the same feature is updated in both the target version and the edit version

2. When the same feature is updated in one version and deleted in the other

For most reconcile operations, no conflicts will be encountered. This is because at most organizations, projects and versions represent distinct geographic areas. If you and your coworkers are editing different parts of the map, it is generally not possible to introduce conflicts. Conflicts usually arise when multiple people are editing features that are in close proximity. There are tools in ArcGIS that allow editors to resolve conflicts interactively during the reconcile process.

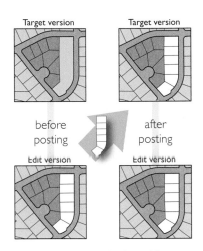

Target version   Target version

before posting   after posting

Edit version   Edit version

During the post, the edited parcel features in the edit version will be pushed to the target version.

Posting versions

Once the two versions have been reconciled with one another and any conflicts have been resolved, the edit version will now contain all the changes from both versions.

As you notice in the graphic to the left, the target version has not changed. The edits that were present in the target version prior to starting the reconcile remain in the target version. The edit version, however, has changed and now includes the edits from both versions. The editor's goal is to get all the edits made in the edit version into the target version. The post-process does exactly this by pushing all the changes from the edit version into the target version.

Once the changes are successfully posted to the target version both versions will be identical. At this point, the editor then has the option to continue to make more edits in the edit version, then perform another reconcile and post-process to synchronize the two versions again, or simply save edits and stop editing the version.

Compressing the geodatabase

The tracking of edits in a version geodatabase environment requires the synchronization of a number of different geodatabase tables. Over time, an actively edited ArcSDE geodatabase accumulates hundreds of thousands of records in these tables, which have become redundant or unnecessary. Compressing the geodatabase is the process of removing these unused records from the versioning schema tables. Actively edited ArcSDE geodatabases should be compressed periodically to maintain optimal performance. Compress can be thought of as an oil change for a geodatabase; the more frequently the geodatabase is used, the more frequently it should be compressed.

Working with nonversioned datasets in an ArcSDE geodatabase

When you create or import a dataset into an ArcSDE geodatabase, it is initially created as a nonversioned dataset. This means that there are no delta tables associated with the dataset and the geodatabase will not track any of the edits made to the dataset. When you make edits to a nonversioned dataset, changes are not tracked and are visible to all editors with permission to view the dataset. As edits are made, instead of going into the delta tables like they do when editing versioned data, they are applied directly to the data source.

You can choose to edit with both versioned or nonversioned datasets in an ArcSDE geodatabase. When you edit a nonversioned dataset, you will have to understand the database behavior with respect to locking, isolation levels, constraints, and triggers to enforce data integrity. Some of the details about database behavior vary depending on your relational database. Consult the ArcGIS help system for details about database behavior for the relational database used for your ArcSDE geodatabase.

You can choose to edit nonversioned datasets only if you are using simple data types such as point, line, and polygon feature classes that do not participate within a data structure such as a geometric network, network dataset, terrain dataset, or topology. Nonversioned datasets are also intended for editing by a single concurrent user. Editing a nonversioned dataset avoids the overhead of managing these extra tables and allows edits made in ArcGIS to be easily picked up by third party applications which typically interact with the database tables directly.

Versioned editing with move edits to base option

Versioned editing with the option to move edits to base provides users with functionalities of both versioned and nonversioned editing. This editing model functions like regular versioned editing except when editing the default version.

When the default version is edited, changes are made directly to the data source and not stored in the delta tables. Also, when versions are reconciled and posted back to the default version, edits in the delta tables are moved directly to the base tables. This allows for a version editing experience while still providing the enhanced third-party application integration available through nonversioned editing.

The move edits to base option is only available for simple data types such as point, line, and polygon feature classes that do not participate within data structures such as a geometric network, network dataset, terrain dataset, or topology.

A common versioning scenario is to use a quality control version as a reconcile and post target for all other versions. This protects the default version from direct editing and unauthorized or unintentional modification. There are three actors in this scenario: Erica is the GIS administrator, Emily and Jonathan are GIS analysts.

Erica, the GIS administrator, must make sure that only valid changes get applied to the default version. Because of this, she decides that the surrogate default workflow is better suited to her organization's strict quality assurance requirements. All other users on the system can see the default version and work with it in a read-only capacity, but they will not be able to edit or post their changes to this version.

Emily needs to make updates to the addressing data. She creates a version off of the quality control version for updating attributes to address points. Jonathan needs to make updates to the water lines and creates another version off of the quality control version for the water utility maintenance project that is under way.

Connecting to each version and making the appropriate edits will isolate the work in each version. Emily and Jonathan spend the rest of the day editing both versions through ArcMap.

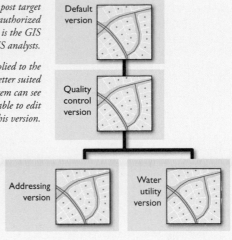

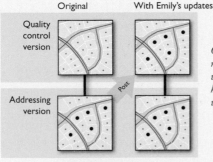

Once the address update project is finished, Emily reconciles the version with the quality control version dealing with any conflicts. She then posts her changes, pushing them into the quality control version.

When finished with the replacement of a water line in the water utility version, Jonathan reconciles that version with the quality control version. The process of reconciling with the quality control version pulls in Emily's address changes that she posted earlier. After Jonathan reconciles, he can then post his changes to the quality control version.

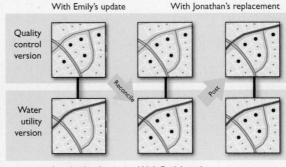

At this point, it is Erica's turn to work with the changes made. She connects to the quality control version and examines the changes Emily and Jonathan have made to the version. After thorough inspection, she reconciles the quality control version with the default version. She then posts edits in the quality control version to the default version.

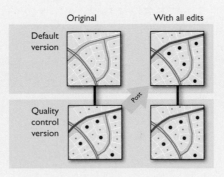

All the changes made to the housing and water projects have been inspected and merged with the default version and are ready to be viewed by the public. Now, Erica can delete the versions that were created for these projects.

Organizations use geodatabase archiving to preserve changes to their data and answer questions such as the following:

"What was the value for a particular attribute at a certain moment in time, such as who owned a particular parcel of land on April 10, 1998?"

"How has a particular feature changed over time, such as the alignment and pavement properties of a section of highway between January 1st, 2010, and March 1st, 2010?"

"How has a spatial area changed over time, such as what are the changes in the roads, buildings, and demographics of a community in the years 2007, 2008, and 2009?"

Versioned dataset in a historic version

When you enable archiving on a versioned dataset in an ArcSDE geodatabase, an additional table is defined called the archive table. The archive table begins with all the rows and fields from the base table of the dataset on which archiving was enabled. The archive table also adds two fields which are used to track the representation of that dataset as it is updated by keeping a timestamp of the time and date when those edits are made.

The versioning architecture of tracking changes in tables as edits are made is further expanded by two powerful technologies: historical archiving and geodatabase replication.

Historical archiving

Geodatabase archiving lets you record and access changes made to data in a versioned geodatabase. It extends the versioning infrastructure by tracking changes and recording the specific date and time when the changes were made. It stores this information in an archive table that is associated with a feature class that has archiving enabled. This archived information is then used in historical queries in ArcGIS such as, "What did this feature look like 2 years ago?" or "Show me all the edits to this feature for the previous tax year."

Geodatabase archiving works by adding an additional type of version called a historical version. As an ArcGIS user, you can connect to either a transactional version (the type of geodatabase version discussed so far in this chapter) or a historical version. The difference between the two types of versions is that the transactional version allows you to perform edits while a historical version is a read-only snapshot of the data in a geodatabase. When connecting to a historical version, you specify a particular date and time. Once connected, you will see a view of the geodatabase just as it appeared on that date.

These are a couple of examples that illustrate how geodatabase archiving can assist you in inspecting the state of your data at moments in time of interest:

- A fire marshal who is monitoring a forest fire can use archiving to map the spread of the flames. By making edits of the fire's perimeter and saving those edits in 20-minute increments, the marshal can build an archive of a fire boundary over time. By color coding the fire's perimeter based on a timestamp field in the archive table, the fire marshal can view the expanding boundary of the fire.

- A GIS staff member in a county assessor's office makes edits over time to a parcel cadastre. Her boss, the tax assessor, needs to view how the parcels appeared at a moment in time such as the end of the year to prepare property tax bills. The tax assessor can connect to the historical archive with that historic marker corresponding to year and examine those records.

Historical archiving is used to present snapshots of features as they change over time, such as these parcels on December 31, July 16, and January 1.

Geodatabase replication

Versioning allows editors of the geodatabase to merge edits between versions using a defined version hierarchy. Geodatabase replication extends this versioning infrastructure by allowing users to merge and synchronize changes across multiple geodatabases. One of the more interesting aspects of versioned geodatabases is their ability to share and synchronize changes. For example, you can broadcast updates from the city GIS department to remote copies of that geodatabase in police, fire, and planning departments. This is called one-way replication and is much like how you can synchronize your media player with new songs using a central media application. In many other cases, users want to exchange changes made within various department geodatabases with their main geodatabase. In this case, two-way replication is used to synchronize these changes both from and to the geodatabases. This is much like how an email account works; you can both get and send email.

This information synchronizing is done by setting properties for how your geodatabases are to be synchronized. This can be scheduled to occur at regular intervals or configured to synchronize only at certain points much like what is done with iTunes® and email.

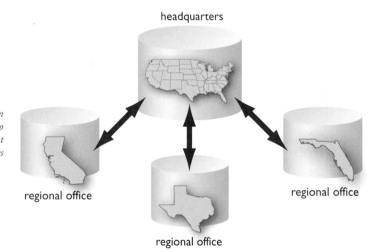

A common pattern of geodatabase replication for an organization's headquarters to disseminate replicas for use throughout regional offices

Many organizational workflows involve the editing of information across many offices and mobile units such as service trucks. Geodatabase replication extends versioning and offers options to support many different types of workflows.

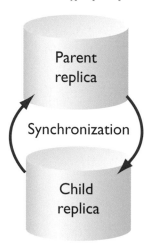

Parent and child replicas are geodatabases that can be edited simultaneously. Datasets in the replicas can be synchronized whenever necessary for your organization's workflow.

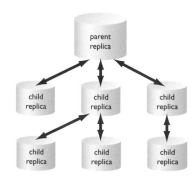

Geodatabase replication is a capability of ArcGIS that allows you to copy data from one geodatabase to another, independently edit datasets in each geodatabase, and then transmit those edits so that the datasets in both geodatabases are synchronized.

Geodatabase replication offers options that support many different types of workflows. It is built on the versioning technology of ArcSDE geodatabases and does not depend on the replication capabilities of the underlaying relational database.

Scenarios for geodatabase replication

These are some situations where geodatabase replication is typically applied:

- Many organizations have several offices throughout the country or around the world. Replication could manage copies of shared datasets that are maintained locally at different facilities. Replication would be used to transfer edits at each facility to other facilities.

- Load balancing is another use of replication. An organization may encounter a performance decrease when serving a central geodatabase due to an increased number of editors. To improve performance, the activities of the data editors could be separated by using a production geodatabase to support edit operations and a publication geodatabase to support data views and Web applications. Replication can be used to push changes from a production geodatabase to the publication geodatabase to keep these two geodatabases in sync.

- Some organizations have field workers creating new data or editing existing data in the field. Replication can be used to copy datasets from a main geodatabase in the office to a geodatabase on a laptop computer. Once edits have been completed in the field, they can be synchronized back to the main geodatabase.

- Organizations sometimes have contracts with companies that provide GIS data processing services. For example, a municipal utility may have a data collection contract with a consulting group to provide field data collection and map those features in a geodatabase. Replication is a good solution for managing the synchronization of datasets in the municipal geodatabase and consultant's geodatabase.

In summary, geodatabase replication is useful for any situation where datasets in geodatabases need to be distributed. It is flexible, versatile, and can be tailored to accommodate many workflows.

Replicas of geodatabases

Geodatabase replication works by managing the relationship between two geodatabases. The original geodatabase is called the parent replica and the geodatabase that contains a copy of the original data is called a child replica. As with versions, you can establish a tree of relationships among replicas.

These are some essential facts to help you understand replicas:

- You can create a replica from the default version or any other version of your ArcSDE geodatabase. If a group in your organization is responsible for maintenance of field inventory, you can create replicas from the maintenance version for distribution to field workers for use on laptop computers in their vehicles.

- You can specify a selection of datasets in a geodatabase version to copy to a replica. It is not necessary to replicate the entire geodatabase. If you are making a replica for a department of an organization, your replica can contain only the datasets of interest to that department.

- You can also apply definition queries, spatial filters, and selection sets to define your replica. For example, if your organization manages an ArcSDE geodatabase for all of Europe, you can create separate replicas for countries such as Sweden, Denmark, and Italy.

Types of replicas

With geodatabase replication, you can choose from three types of replicas: Checkout/check-in, two-way, and one-way.

Checkout/check-in replication allows you to copy data to another geodatabase, edit that data, and when you're done, transfer those edits back to your original geodatabase. With this type of replication, you can perform the check-in operation only once. If you need to perform more edits once you've checked in the data, you must create another checkout replica.

One-way replication lets your organization send changes many times from a parent replica to a child replica, or from a child to a parent. One-way replication is useful for a situation where you have a production geodatabase with continual edits and a publication geodatabase optimized for map making and Web mapping services.

Two-way replication lets you send changes back and forth between replicas many times without having to re-create the replica. Changes can be sent from either the parent replica to the child replica, or from the child replica to the parent replica, or you can send changes in both directions at the same time. Two-way replication is useful for situations where two groups need to share data that they both need to edit. For example, if an organization is working with a consulting group, two-way replication allows the simultaneous editing of both replicas and their synchronization.

Connections for synchronization

You can create and synchronize replicas in several types of connectivity environments. Replicas can be created and synchronized over local area network connections, Web services, or published securely through the Internet. If two geodatabases cannot access each other through a network, then you can use XML files to create or synchronize replicas and transfer the XML file through email, FTP, or a DVD.

Checkout/check-in replicas

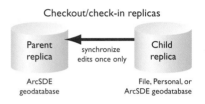

One-way replicas

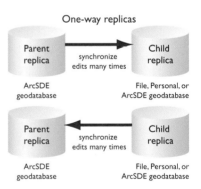

Two-way replicas

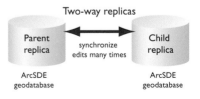

When you apply geodatabase replication, there are two steps to follow.

The first step is to identify and prepare the data for replication. You can apply schema changes across replicas after replica creation, but this requires some work and not all schema changes are supported. Consult the ArcGIS help system for details on preparing data for replication and working with schema changes.

The second step is to create the replicas and determine a synchronization strategy. The strategy depends on your use case for replication; six common use cases are discussed in this topic.

Replication for mobile users

Mobile users can take ArcGIS Desktop software to the field on a laptop computer and perform edits on a geodatabase replica. For this use case, you will probably want two-way replication. With two-way replication, you can synchronize multiple times from the field to the office or from the office to the field. You can also use checkout/check-in replicas, but this will require you to re-create a new replica after each synchronization.

In this use case, laptop computers are typically returned to the office at the end of the workday. Synchronization can take place overnight and the laptops can be ready for another day's field editing work in the morning.

With multiple mobile users, there are two possible strategies for synchronization: two-way synchronization can be done with each laptop in turn, or all laptops send their updates to the office geodatabase and all laptops receive back the collective updates from the office geodatabase.

Replication for multiple offices

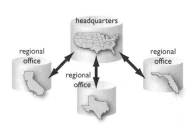

When there are multiple offices served by a central office, the central office can serve as a hub from which replicas are created and distributed to the other offices. Once replicas are created, each office has its own copy which can be worked with directly. Offices then share changes with one another by synchronizing with the central office.

For this use case, two-way replicas are required to share changes from one satellite office with other offices. Synchronization is typically done on a regular schedule with the central office responsible for resolving conflicts and redistributing replicas.

Replication for multilevel organizations

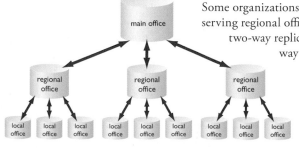

Some organizations have offices at several levels, such as a main office serving regional offices which in turn serve local offices. For this use case, two-way replicas should be used. The main office would create two-way replicas for each of the regional offices and the regional offices would create two-way replicas for the local offices.

Most of the edits are probably done at the local offices. Synchronization would proceed by sending changes from the local offices to their regional office, as well as receiving changes back from the regional office. Once the regional office receives edits from the local offices, then those changes are passed up to the main office. Synchronization can be scheduled so that iterations through this synchronization pattern can be repeated on a regular basis.

Replication for working with contractors

You can use replication to distribute data to contractors. You can replicate just the data that the contractor requires to perform the desired data edits. As work progresses, the contractor can send updates to the main office.

If you want to accept changes from the contractor multiple times, create a two-way replica. If you want to accept changes only once, the job is complete, then create a checkout/check-in replica.

Replication for production and publication geodatabases

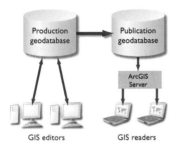

Some organizations serving geographic data on the Internet distribute the load by configuring a production geodatabase for GIS editors and a publication geodatabase for GIS readers.

For this use case, one-way replicas work well. Since the flow of edits goes only from the GIS editors to the GIS readers, changes need only be sent from the production geodatabase to the publication geodatabase. You can also replicate only the data required by the GIS readers.

Synchronization should be scheduled at a time when GIS readers are least likely to access the data in the publication geodatabase, such as the middle of the night. With one-way replicas, there are no conflicts to be resolved.

Replication for multigroup geodatabases

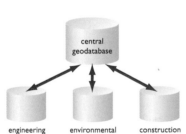

Sometimes, organizations distribute data on the basis of functional responsibility instead of geographic areas, such as geodatabases for engineering, environmental, and construction groups.

For this use case, a replica is created for each group. Two-way replication is used so updates can be exchanged between each project group and the central geodatabase on a continual basis. Scheduling synchronization for this use case is simpler than synchronization for multiple offices of multilevel organizations because no edit conflicts should occur. So synchronization can be done at any time when each group is done with a project.

Transmission map detail of northern New Mexico.

Public Service of New Mexico (PNM) provides electric power to a large service area spanning 55,000 square miles and 3,000 miles of high voltage transmission lines. Besides serving a large area, PNM faces requirements in keeping compliant with environmental regulations, protecting archaeological sites in the vicinity of transmission equipment, and effective response to electrical outages—all with a relatively small transmission field crew.

To serve these requirements, PNM developed a geographic information system in partnership with Power Engineers, Inc. The system is built on ArcGIS and was designed in close coordination with the linemen who use it on a daily basis.

PNM uses ArcGIS to manage the editing, analysis, and display of large amounts of geographic information: the electrical transmission infrastructure, landownership, rights-of-way and access, archaeological sites, and environmental habitats.

Analysts and field crew can perform these and other activities:

- Find the fastest route to any facility on the transmission system while avoiding environmental or culturally sensitive sites and respecting land ownership rights.

- Provide field personnel with engineering data, material lists, maintenance history, plan and profile drawings, and any other information needed to complete a job.

- Enter inspection information, including digital photographs of the equipment that requires maintenance.

- Manage transmission asset inventory so that reports of structures and line conductors can be provided to local tax jurisdictions.

PNM personnel use geodatabase versions and replicas to manage the flow of information to and from the field, as well as analysis done at headquarters.

There are two main user groups of ArcGIS within PNM: the linemen who perform the daily inspection and maintenance of the transmission system and the support engineers, biologists, archaeologists, and other analysts who manage the information needed to meet PNM's mandates for customer service, environmental responsibility, and regulatory compliance.

Each service truck has a computer operating ArcGIS. A geodatabase replica is created for every truck and is used for two-way updates of information about transmission equipment.

Each group of support staff at headquarters also has a geodatabase version. Besides isolating edit transactions within each group, until those edits are posted to the default version, the group versions are used to protect the confidentiality of sensitive data, such as archaeological sites. Pilfering of pots and other Native American artifacts is a problem in the southwestern United States and PNM uses versions to restrict access to sensitive data.

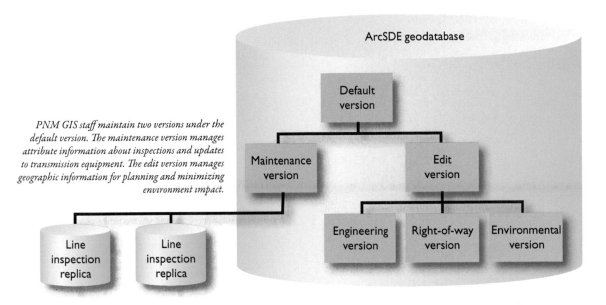

ArcSDE geodatabase

Default version

PNM GIS staff maintain two versions under the default version. The maintenance version manages attribute information about inspections and updates to transmission equipment. The edit version manages geographic information for planning and minimizing environment impact.

Maintenance version

Edit version

Engineering version

Right-of-way version

Environmental version

Line inspection replica

Line inspection replica

The maintenance version has several replicas which are installed on laptops in PNM maintenance trucks.

A line inspector drives a truck along a transmission line performing routine inspection at each tower. At one tower, he discovers damage to a string of insulator bells that attach a transmission line to the structure. He photographs the condition and enters field notes to his line inspection replica on the truck laptop computer. He updates attributes for that tower in the Support Structure feature class, records the severity of the damage, and transfers the photograph from the digital camera to the computer with a link to the structure feature. He returns to headquarters and posts the update to the maintenance version.

A transmission line crew downloads a line maintenance replica and drives to the structure to perform the required maintenance. As the work is done, this crew makes notes on their replica in the truck computer. When they return, they post updates from their replica to the maintenance version.

A GIS supervisor periodically reviews the changes made to the maintenance version and posts updates to the default version.

Several groups perform edits and updates to the geodatabase through the edit version. Each group has a version.

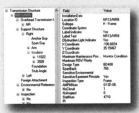

The engineering version is used for power analysis of the transmission system. This screen shows details of the system for study by the engineer.

The right-of-way version is used to manage utility easements and roadway access for servicing transmission equipment. The structures shown with red symbols have associated restrictions.

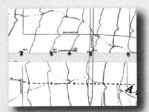

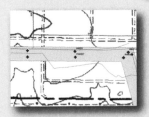

The environmental version is used to map archaeological sites and sensitive habitats for endangered species. This data is used to warn field crews to avoid certain locations. Specific information about archaeological sites and habitats is kept confidential.

Geoprocessing with models and scripts

Geoprocessing is how you compute with geographic data. GIS users in hydrology analyze surface models to derive stream lines and study the accumulation of rainfall. GIS users in natural resource management analyze environmental factors such as soil and vegetation types to protect habitats. GIS users in urban planning study which areas are best suited for developing affordable housing. Geoprocessing models and scripts are tools to perform geographic analysis and an important component of GIS data models.

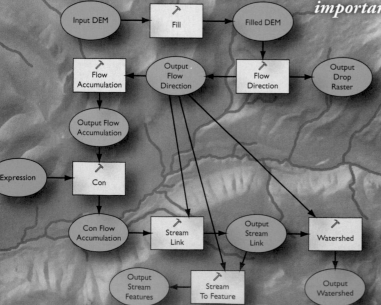

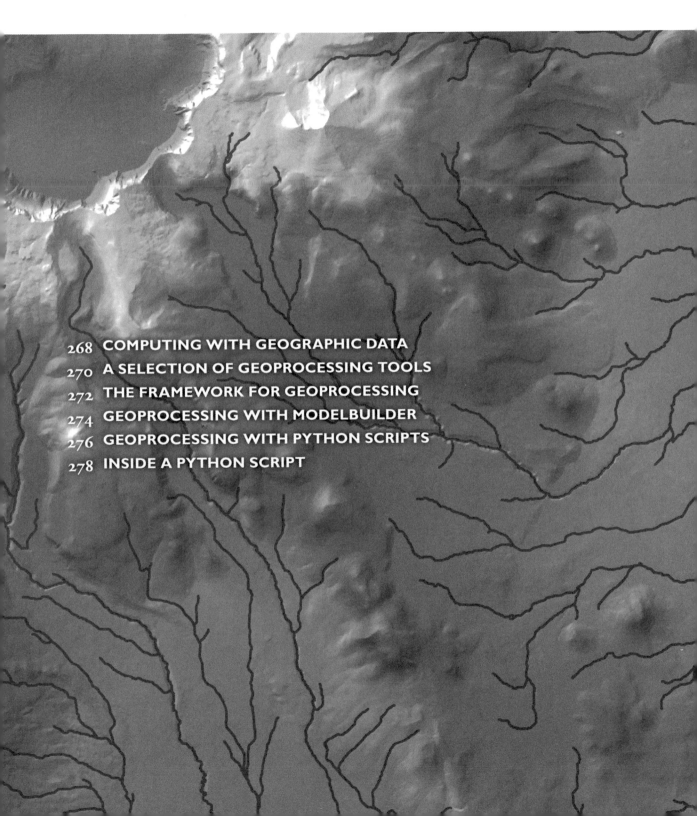

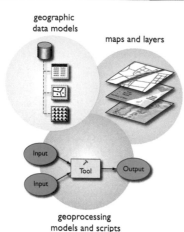

A functional geographic information system for an application is built by designing and creating a geographic data model, map and layer templates, and geoprocessing models and scripts.

Geographic data models are developed as templates for the data loaded into a geodatabase. Geographic data models are designed to capture the attributes of geographic features and systems that are important to your application.

Map and layer templates are used to apply cartographic consistency to the presentation of the data in your geodatabase. Maps use standardized symbols and visualization methods to communicate geographic relationships.

Geoprocessing models are created to handle the spatial analysis of your geographic data and to automate data management tasks.

Geoprocessing is one of the three critical components of a geographic information system, along with geographic data models and maps.

- Geographic data models are designed to store the information required for your application and to ensure the integrity of the data you store.

- Maps and layers let you visualize and edit your data and explore relationships among geographic entities.

- Geoprocessing models and scripts are used for computing with geographic data and automating many repetitive tasks.

Geoprocessing is all about applying tools to existing data and creating new data for spatial analysis and data management. You can apply ArcGIS geoprocessing tools to execute geographic operations in several ways: by interactively executing a tool through a dialog box, by typing in a tool name and parameter values in a command line, by creating and running a geoprocessing model in a graphical environment called ModelBuilder, and by building a script that can automate the execution of many tools in a sequence.

ArcGIS supplies many hundreds of geoprocessing tools for practically every conceivable geographic operation. These tools perform operations on geographic data such as extracting and overlaying data, changing map projections, adding columns to a table, calculating optimal routes, and performing surface analysis such as slope and visibility.

Overview of geoprocessing

Geoprocessing operations transform data from one form to another.

Each geoprocessing operation starts with one (or more) input datasets and parameters and applies a geoprocessing tool to produce a new dataset.

Geoprocessing models and scripts are built by chaining multiple tools, each of which performs a discrete geographic operation on one or more input datasets to create a new dataset. For each tool in the model, you can specify all the parameters for that process, such as distances, tolerances, and inputs —just as you would if you ran it interactively.

There are two broad uses for building models and scripts.

The first broad use is to perform complex spatial analysis in a well-defined and repeatable way. Many organizations need to analyze problems in a geographic context that involve the interaction of multiple factors. An example is analyzing weather patterns and terrain to predict good locations for siting a wind farm.

The second broad use is to automate repetitive data workflows that many organizations perform, such as converting data from one form to another, manipulating data and relationships, and managing spatial references.

Scenarios for geoprocessing

To understand how geoprocessing can be used for spatial analysis, it's helpful to review several scenarios that a planner at a local government my encounter. Each of these involve the use of multiple geoprocessing tools in a sequence and these steps can be stored in a geoprocessing model or script.

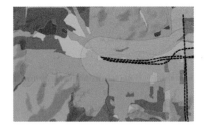

Buffered areas around proposed roads are overlaid on top of vegetation polygons to study the impact on an endangered songbird habitat.

A new set of roads is being proposed in your local area to handle population growth. But one characteristic of this area is the presence of an endangered songbird. You want to ensure that these planned roads do not impact the critical habitats for the songbird. In consultation with a wildlife biologist, you identify the important environmental factors for the survival of the songbird, such as vegetative types, elevation range, and whether slopes are oriented to the north or south. You then build a geoprocessing model that performs the appropriate geoprocessing tools on the input datasets, and then determine if any optimum habitats are within a buffer zone around the proposed roads. With this information, you submit an environmental report stating whether or not the planned roads endanger the songbird.

Another task you may have as a planner is to analyze which roads may be flooded in a major storm. Given an estimated rainfall over an area, you run a hydrological model that factors in the water shed boundary, local terrain, and soil impermeability and from those factors simulates a storm event. The output from the model are estimated floodplain areas which you then overlay on a road network to identify areas where motorists are at greatest risk of sudden flood. Once these areas of flood potential are identified, an engineer can mitigate the risk by designing culverts or bridges as needed.

Estimated floodplain areas are overlaid on a terrain dataset to locate roadways which may be under water during a major rain event.

A mundane but essential aspect of your job as a planner is to periodically import data from another agency and manipulate this data so that it is in a form useful to your agency. Because this activity happens repeatedly, you build a geoprocessing model that imports the data, changes the map projection to your standard, and adds fields required by your organization.

Benefits of applying geoprocessing models and scripts

The scenarios just discussed can all be performed interactively with ArcGIS. But performing these operations step-by-step can be time consuming and prone to error. Building geoprocessing models and scripts are valuable for any analysis that is repeated more than once and has these benefits:

- Models and scripts let you repeat an analysis many times with modified parameter values or different input datasets.

- Models and scripts let you share a process with other people and they can perform the same analysis at their site.

- Models and scripts help formalize a process of spatial analysis. The assumptions in your spatial analysis are documented and can be modified to reflect improved scientific understanding of a geographic problem.

Models and scripts can be included within a geodatabase. You should think of geoprocessing models and scripts as an advanced yet essential component of your geographic information system.

There are two parts to geoprocessing in ArcGIS:

— the many geoprocessing tools that are available for spatial analysis and data management and

— the framework for applying these tools within geoprocessing models and scripts.

This topic reviews some examples of geoprocessing tools.

The tool examples presented here are not necessarily the most commonly used, but are shown to illustrate the range of tool operations. Three tools below on the left show geoprocessing operations on raster datasets. Three tools below on the right show geoprocessing operations on digital elevation models (rasters with elevation values). The tools on the facing page show geoprocessing operations on feature classes.

There are many hundreds of geoprocessing tools that perform operations on every data structure supported in ArcGIS. There are tools to create new geodatabases, tables, feature classes, and rasters. There are tools to import, manipulate, convert, and process data. There are tools designed for geocoding, surface analysis, geostatistical analysis, network analysis, and managing geodatabase versions. It is beyond the scope of this book to summarize all of these tools; consult the ArcGIS help system for detailed descriptions of each tool.

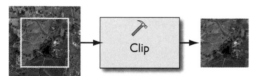

The clip tool uses a polygon boundary to extract a portion of a raster to create a new raster.

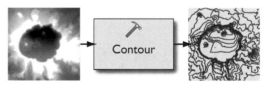

The contour tool creates a line feature class with contours interpolated from a digital elevation model.

The mosaic tool combines a number of rasters into one merged raster.

The hillshade tool creates a raster with gray values depicting shaded relief from a digital elevation model.

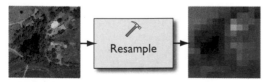

The resample tool creates a new raster from an input raster with a different cell size.

The aspect tool uses a digital elevation model to create a raster showing the orientation of slope (aspect) with colors.

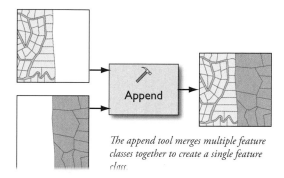

The append tool merges multiple feature classes together to create a single feature class.

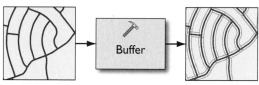

The buffer tool is used for calculating proximity. This tool creates a new feature class with buffer polygons around point, line, or polygon features.

The dissolve tool is used for generalizing features. This tool combines features that share common attributes.

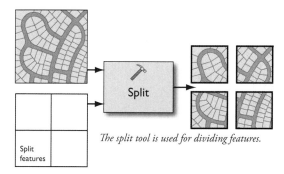

The split tool is used for dividing features.

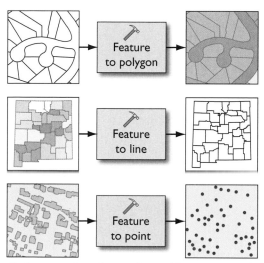

The feature to polygon, feature to line, and feature to point tools change features from one shape type to another. Points are created at the centroids of polygons or midpoints of lines.

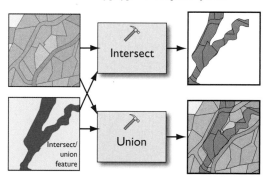

The intersect tool creates intersecting features common to both feature classes. The union tool intersects features and combines them.

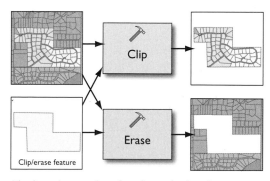

The clip tool uses a polygon boundary to divide and extract features. The erase tool uses an erase boundary to divide and erase features.

The previous chapter topic presented a number of example geoprocessing tools to give you a sense of the many geoprocessing tools available in ArcGIS. This chapter topic introduces the framework for working with geoprocessing tools.

There are four ways of running geoprocessing tools in ArcGIS.

- You can run a geoprocessing tool through its dialog box in ArcGIS. Inside a tool's dialog box, you simply enter parameter values and then run the tool. Online help is available to guide you through the options. Often, a tool has required and optional parameters.

- You can run a geoprocessing tool through a command line interface in ArcGIS. The tool name and its parameter values are typed in as a text string. The tool executes when you press Enter on the command line.

- You can build a model of your geoprocessing workflow by stringing many geoprocessing tools together. For each tool in a model, you will supply parameters through the same dialog box as when you run the geoprocessing tool interactively. This activity is done in the Model-Builder window, accessible from ArcGIS applications.

- You can create a script in one of several scripting languages and then run the script from a toolbox.

The first two ways—dialog box entry and command line entry—directly apply geoprocessing tools. These methods are appropriate if your operation is simple and is not part of a process which you repeat. Using a dialog box is a good way to familiarize yourself with geoprocessing tools.

The second two ways—models and scripts—let you capture a workflow by combining multiple geoprocessing tools. Models and scripts are recommended if workflows have multiple steps and are repeated. Once created, models and scripts can be edited and reused.

An easy way to start building scripts is to first build a model. Once you've created a model, you can export it to any supported scripting language and add more programming logic.

Create models if the simple graphical user interface of the ModelBuilder window appeals to you. Develop scripts if you are comfortable with programming and want to exercise the full capability of a scripting language. Scripts also give you the flexibility of running a process outside an ArcGIS application.

Toolboxes and tools

Toolboxes are where you access tools and store your models and scripts. Toolboxes contain toolsets, tools, models, and scripts. Toolsets can also contain toolsets, tools, models, and scripts. You access toolboxes from the ArcToolbox window or the ArcCatalog tree.

You should think of models and scripts as custom tools which augment the system tools provided in ArcGIS. Just as tools have dialog boxes for entering parameters, you can create dialog boxes for the parameters you define for models and scripts. Parameters can specify input datasets, output datasets, and values such as distances.

The tools executed within a model can be system tools, other models, or scripts. The tools executed within a script can be system tools or other scripts. Models are stored in toolboxes. Scripts are stored in toolboxes and also as text files in the file system. Custom toolboxes are stored either inside a geodatabase or as a toolbox file (.tbx extension) in the file system.

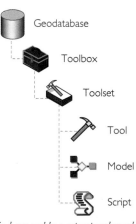

Geodatabase

Toolbox

Toolset

Tool

Model

Script

You can find any tool by navigating through toolboxes and toolsets, but the easiest way to find a system tool in the ArcToolbox window is by searching for keywords.

Imagine that you want to identify a set of gas stations that are located within 300 meters of highways. This simple analysis is done by combining two geoprocessing tools. Below, you can see how this spatial analysis task is done with both a simple Python script and a geoprocessing model built in ModelBuilder. You can choose which environment you prefer, the graphical user interface of ModelBuilder or the programming environment of a scripting language such as Python.

This Python code is simplified to show how to perform this analysis with just four lines of code. (The lines beginning with '#' are comment lines.)

The first line of code imports ArcPy so that geoprocessing tools in ArcGIS can be executed.

Note that this code is simplified to the point where dataset locations are hard-coded and no error handling is done. A more robust script would handle input parameters and test for error conditions.

```
# Import ArcPy to enable ArcGIS geoprocessing
import arcpy

# Set the workspace environment
arcpy.env.workspace = "c:/City/Transportation.gdb"

# Buffer highways with buffer distance of 300 meters.
arcpy.Buffer_analysis("highways", "Highway_Buffers", 300)

# Find gas station points that intersect highway buffer
arcpy.Intersect_analysis(["GasStation", "Highway_Buffers"],
                "GasStationsNearFreeways")
```

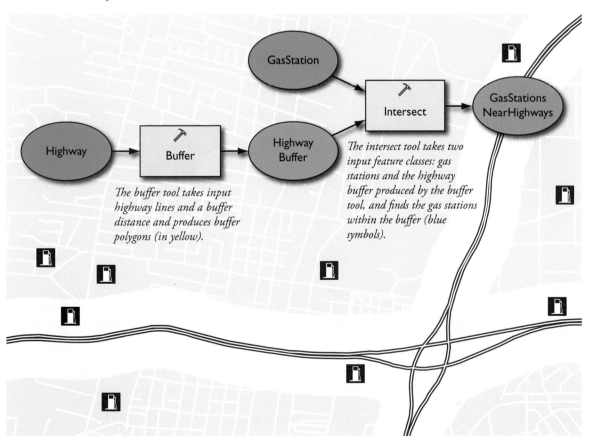

The buffer tool takes input highway lines and a buffer distance and produces buffer polygons (in yellow).

The intersect tool takes two input feature classes: gas stations and the highway buffer produced by the buffer tool, and finds the gas stations within the buffer (blue symbols).

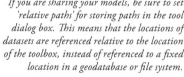

The ModelBuilder window in ArcGIS is an easy way to automate any spatial analysis that has multiple steps. To build a model, you simply add elements and connect them just as you would draw a flow chart. No programming knowledge is necessary and models can perform powerful geographic analysis.

Parts of a model

A model is a collection of processes. Each process is a combination of an input variable, tool, and output variable.

Elements for tools appear as yellow rectangles once all parameters have been defined. Tools can be ArcGIS system tools, other models, or scripts. Once you place a tool in a model, you can define the parameters for its input dataset, output dataset, and other values through the tool's dialog box. For system tools, this is the same dialog box as you would see if you ran that tool interactively. Models and scripts have custom dialog boxes for their parameters.

There are three types of tool elements in a model. They are identified by their icons as system tools, model tools, or script tools.

Elements for variables are drawn as ovals. Once variable elements are defined as tool parameters, input variables are drawn as blue ovals and output variables are drawn as green ovals. Variable elements can represent datasets or other values such as distances.

Within a model, each process has three states: not ready to run, ready to run, and has been run. When you first place an element in a model, it appears as a white shape. The element changes to blue for an input element, yellow for a tool element, or green for an output element once the parameters for the connected tool element have been defined.

If you are sharing your models, be sure to set 'relative paths' for storing paths in the tool dialog box. This means that the locations of datasets are referenced relative to the location of the toolbox, instead of referenced to a fixed location in a geodatabase or file system.

Once each element has been executed within a model, that element shape acquires a drop shadow. This lets you know the current status of the execution of your model.

Connector lines define how the input and output elements are associated with tool elements. Labels are optional, but useful for documenting your model.

Looping through inputs in a model

Another element you can place in a model is an iterator. An iterator lets you run a model multiple times for multiple inputs.

There are a number of types of iterators. Some examples are iterators that run a fixed number of times, iterators that run until a condition is true or false, iterators that run for each feature in a feature class, iterators that run for each row in a table, iterators that run for each field value in a table, iterators that run for each dataset in a geodatabase, iterators that run for each raster in a geodatabase, iterators that run for each feature class in a feature dataset, iterators that run for each table in a geodatabase, and more.

Iterators are an advanced model element that lets you apply a model to multiple inputs.

One limitation of iterators is that a model can have only one iterator. If you need to perform multiple iterations, you can achieve this by nesting models; that is, one model invoking another model tool containing an iterator.

Define model tool parameters

Below is a sample model that processes an input digital elevation model (a raster with elevation values) and creates output stream and watershed features. This model was developed at ESRI for the common hydrological task of interpolating streams and watersheds from a surface.

You will see that some input and output elements have a letter 'P' beside them. This indicates that these elements are defined as model parameters. When executed as a tool, the user will first encounter a custom dialog box and set parameters for the input datasets, output datasets, and other values.

Typically, when you first create and develop your model, you will define fixed parameters for datasets and values until your model is well tested. Once you are ready to apply the model to other datasets or share it with other users, then you can define input and output elements as model parameters. Once you do so, a custom dialog box is generated on-the-fly when the model is run, prompting the user for datasets and values.

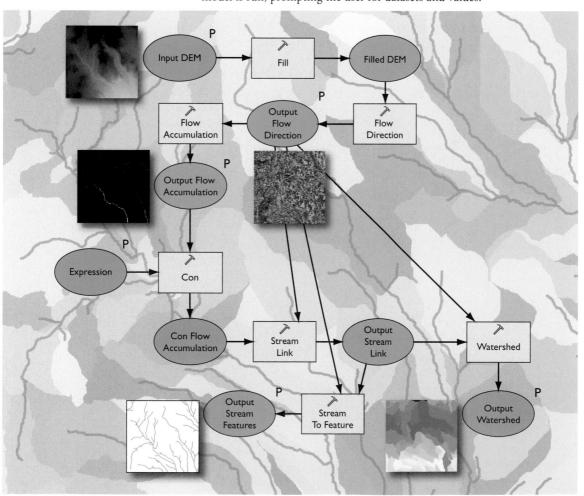

Python is the recommended scripting language for automating workflows and spatial analysis with ArcGIS. With Python, you can perform simple or complex processes. Python is ideal for creating generic processes that are easy to reuse and distribute to other people.

You can perform geoprocessing with Python in two modes, as stand-alone scripts or as script tools in ArcGIS:

- Script tools run within the ArcGIS applications and interacts with those applications by sending messages, setting environment values, and adding output datasets to a map. A script tool has full membership in the geoprocessing framework and work just like any other system tools in ArcToolbox.

- Stand-alone scripts can be run outside of ArcGIS from the Windows command window, like any .exe or .bat file. Stand-alone scripts can be scheduled to run anytime and can call any geoprocessing tool.

Python and ArcGIS geoprocessing tools

All geoprocessing tools, as well as a wide variety of useful functions for interrogating GIS data, are available in the ArcPy site-package. A site-package is Python's term for a library that adds additional functions to Python, and ArcPy is how GIS functions are added to Python. ArcPy is installed with ArcGIS.

Some example functions in ArcPy are ListFeatureClasses, to list feature classes in a workspace (geodatabase or folder); Describe, to retrieve information about a feature class; and SearchCursor, to get field values from a table.

You can also set environment settings for ArcGIS, such as the workspace that datasets will be read from or written to, the output coordinate system of new datasets, or cell size of a new raster dataset. This sample code starts by importing arcpy, sets the workspace to a geodatabase, and performs a geoprocessing tool, Intersect.

```
# import ArcPy
import arcpy

# set the workspace
arcpy.env.workspace = "c:/project/GISdata.gdb"

# apply Intersect geoprocessing tool
arcpy.Intersect_analysis(["Roads", "urban_area"],
                         "urban_roads", 5, "join")
```

Handling messages and errors with Python

Whenever you execute a geoprocessing tool in Python, one of three types of messages are produced. Informative messages state that the tool has completed successfully. Warning messages caution about potential problems, such as the creation of an empty output. Error messages inform you that the geoprocessing tool was not successfully executed and states the condition that caused the error.

Use the GetMessages function to display messages. You can use this function to receive only informative messages, only warning messages, only error messages, or all messages.

You should handle errors in Python with the Try... Except... construct. In Python code, it appears as:

```
Try:
  # do something
Except:
  # A failure has occurred, do an action to recover
```

The Try... Except... construct should include the GetMessages function to display informative, warning, and error messages as appropriate.

Running geoprocessing tools on multiple datasets

A common scenario for a script is to run a geoprocessing operation multiple times on a each dataset in a geodataset or folder. ArcPy has a set of useful functions to identify datasets. Here's a simple fragment of code that prints all the feature classes in a workspace:

```
# Get a list of feature classes and print them one at a time
fcList = arcpy.ListFeatureClasses ("*", "polygon")
for fc in fcList:
  print fc
```

Adding logic to a Python script

Most Python scripts contain logic to test for conditions and handle them as necessary. Conditional branching is done through the If.. Else... construct in Python. The Describe function is used to return an object for characteristics of a dataset. Here's a simple code fragment that tests for type of feature class:

```
# Describe a feature class
d = arcpy.Describe("c:/project/GISdata.gdb/rivers")
# Branch based on a input's shapeType property
if d.shapeType == "Polyline":
  print "This is a line feature class"
else:
  print "This is not a line feature class"
```

Receiving arguments to a script

Many scripts will receive arguments, such as names of datasets, so they can run in any setting. Rather than hard-coding the names of datasets and workspaces, you can pass them as arguments to a script. This is done with the GetParameterAsText function.

This simple script takes an input and output feature class name and performs a clip operation on one feature class against another:

```
# Get an input and output feature class as a text string
inputFC = arcpy.GetParameterAsText(0)
outputFC = arcpy.GetParameterAsText(1)
clipFC = "c:/base/GISdata.gdb/boundary"
arcpy.Clip_analysis(inputFC, clipFC, outputFC)
```

This topic introduces some key concepts of working with Python and ArcGIS. Several other resources are available to help you apply Python to geoprocessing.

For scripting syntax and tool documentation, go to the ArcGIS desktop help and read the topics on geoprocessing.

For sample scripts and models, go to the Geoprocessing resource center at http://resources.arcgis.com.

Get a good reference book on Python. One popular book is Learning Python *by Mark Lutz.*

Learn about Python from the Python Organization Web site: www.python.org

A good way to understand how to develop Python scripts in ArcGIS is to study a sample script. The script on the facing page takes all of the feature classes it finds in a workspace, clips them using another feature class with a clip polygon, and outputs the resulting clipped features into a set of feature classes in a workspace.

To enable ArcGIS geoprocessing, this script first imports the arcpy site-package. The workspace is identified as well as the feature class with the clip feature. From a list of all the shapefiles in the workspace, a set of new feature classes are created in a geodatabase.

This script demonstrates how to work with input parameters. Four parameters are defined for the input workspace, the clip feature class, the output workspace, the cluster tolerance. While the illustration below shows shapefiles being clipped into new feature classes in a geodatabase, this script is versatile and can handle feature classes in a geodatabase or shapefiles for input and output.

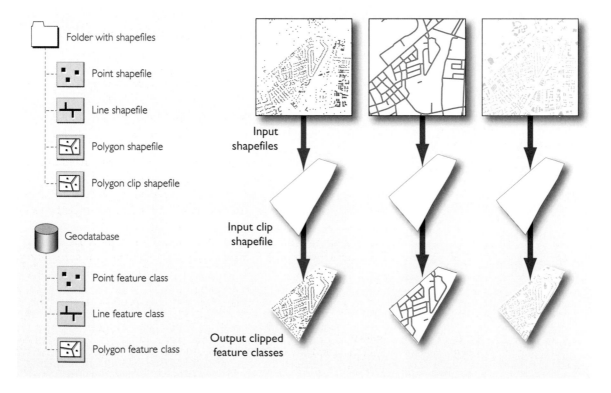

Folder with shapefiles
- Point shapefile
- Line shapefile
- Polygon shapefile
- Polygon clip shapefile

Geodatabase
- Point feature class
- Line feature class
- Polygon feature class

Input shapefiles

Input clip shapefile

Output clipped feature classes

```
# Script Name: Clip Multiple Feature Classes
# Description: Clips one or more feature classes from a
#              workspace and places the clipped feature
#              classes into another workspace.

# Import ArcPy site-package and os modules
#
import arcpy
import os

# Set the input workspace
#
arcpy.env.workspace = arcpy.GetParameterAsText(0)

# Set the clip featureclass
#
clipFeatures = arcpy.GetParameterAsText(1)

# Set the output workspace
#
outWorkspace = arcpy.GetParameterAsText(2)

# Set the cluster tolerance
#
clusterTolerance = arcpy.GetParameterAsText(3)

try:
    # Get a list of the feature classes in the input workspace
    #
    fcs = arcpy.ListFeatureClasses()

    for fc in fcs:
        # Validate the new feature class name for the output workspace.
        #
        featureClassName = arcpy.ValidateTableName(fc, outWorkspace)
        outFeatureClass = os.path.join(outWorkspace, featureClassName)

        # Clip each feature class in the list with the clip feature class.
        # Do not clip the clipFeatures, it may be in the same workspace.
        #
        if fc <> os.path.basename(clipFeatures):
            arcpy.Clip_analysis(fc, clipFeatures, outFeatureClass,
                                clusterTolerance)
except:
    arcpy.AddMessage(arcpy.GetMessages(2))
    print arcpy.GetMessages(2)
```

A 'workspace' refers to a geodatabase, a folder with shapefiles, or any container of geographic data which is accessible by ArcGIS.

The first input parameter is the workspace with the input feature classes.

The second input parameter is a feature class with clip features, which can be a shapefile or a feature class in a geodatabase.

The third parameter is an output workspace in which to place the resulting clipped feature classes.

The fourth parameter is a tolerance used by the Clip tool.

The try: except: code block executes the core logic of the script. If the input parameters are valid, then the code will execute the clip operation on each feature class.
Note that the list of feature classes is extracted from the workspace by using the ListFeatureClasses function.

This code is executed only in the event of an error in the code within the try: construct.

Putting geodatabases to work in your GIS

Throughout this book, you've learned the concepts behind the geodatabase and how the datasets housed within create modern geographical information systems. You've looked at the various ways ArcGIS takes real-world entities and models them using thematic layers stored as rows and features in a geodatabase. Geographic datasets model behavior and relationships that match the complexity of their real world counterparts. They are overlaid on maps that cover a common geographic area. These maps are enhanced with terrains and raster imagery used as basemap layers or as surface models for display and analysis.

GIS analysts work with datasets on these maps in single and multi-user editing environments. Editors can make updates to the datasets and create models and scripts for geographic processing, analysis and automated editing tasks. All of these elements combine to model real-world systems such as water distribution, utility and transportation networks, location and routing services, and GIS solutions in many other industries. Throughout the 11 chapters of this book, you've been presented with examples of these solutions in the form of case studies from the user community.

But how do you take these elements and create the thematic layers and maps like the ones found in this book? How do you apply maps and layers to create these solutions and publish and share information with your organization and others? How do you put the geodatabase to work for your GIS?

The best way to begin to apply what you have learned is to engage with the larger ArcGIS community. Find users with similar interests and challenges. Share your knowledge and learn from the expertise and experience of others.

The ArcGIS Resource Center (http://resources.arcgis.com) is a Web-based portal that provides access to dynamic Web help, blogs and community pages, support information, downloadable templates to help you get started applying ArcGIS, and much more. The ArcGIS Resource Center is a place where you can get connected with users in your user community and elsewhere and find useful and up-to-date information.

ArcGIS.com (http://www.arcgis.com) is another way to find and use content from ESRI and other authoritative sources. ArcGIS.com is an online GIS where you can find data, work with it, share it, and build communities. You can find detailed, multi-scale basemaps and other content to use in your map, search open Web and GIS servers for additional content, create mashups by adding content published by others, and save your map to share with other users. At ArcGIS.com you can also find, create and share useful and interesting applications and solutions. Or you can build on maps and applications shared by the community. This is an essential part of using ArcGIS.

The themes and concepts found throughout this book are alive and in practice on both ArcGIS.com and the ArcGIS Resource Center. These online services are taking what ESRI has been working on from the beginning and creating a paradigm shift towards making GIS accessible to everyone in an easy, innovative and open online experience.

This book is your foundation for knowledge on how to build a GIS solution using the geodatabase and its geographic datasets. The next step is to visit the ArcGIS Resource Center and ArcGIS.com, get connected with other users, content and ideas, and apply the concepts you've learned to begin modeling your world.

Image and data credits

Chapter 5

Page 142–145, courtesy of U.S. Geological Survey and Tele Atlas North America
Page 146, courtesy of Maryland Department of Transportation State Highway Administration
Page 146, courtesy of Parsons Brinckerhoff, Inc.
Page 146, courtesy of Parsons Brinckerhoff, Inc.
Page 147, courtesy of TGS–NOPEC Geophysical Company
Page 147, courtesy of M.J. Harden Associates, Inc.
Page 147, courtesy of Center for Research in Water Resources
Page 149, courtesy of Tele Atlas North America
Page 152–153, courtesy of Tele Atlas North America
Page 155–156, courtesy of U.S. Geological Survey and Tele Atlas North America

Chapter 6

Page 164–165, courtesy of City of Santa Fe
Page 166, courtesy of Tele Atlas North America
Page 169, courtesy of Tele Atlas North America
Page 173, from ESRI Data & Maps 2008, courtesy of Tele Atlas North America, U.S. Census, ESRI and City of Santa Fe
Page 175, from ESRI Data & Maps 2008, courtesy of Tele Atlas North America, U.S. Census, ArcUSA, ESRI and City of Santa Fe
Page 176, courtesy of City of Santa Fe
Page 179–181, courtesy of City of Santa Fe

Chapter 7

Page 182–183, courtesy of City of Santa Fe
Page 185, © 2009 Oregon Metro
Page 186–188, Landsat imagery courtesy of NASA Goddard Space Flight Center and U.S. Geological Survey
Page 188, courtesy of U.S. Geological Survey
Page 188, courtesy of GeoEYE
Page 188, courtesy of City of Santa Fe
Page 189, courtesy of U.S. Geological Survey
Page 189, Landsat imagery courtesy of NASA Goddard Space Flight Center and U.S. Geological Survey
Page 190, courtesy of U.S. Geological Survey
Page 190, courtesy of City of Santa Fe
Page 190–191, courtesy of U.S. Geological Survey
Page 191, from ESRI Data & Maps 2008, courtesy of NASA, NGA, USGS EROS, and ESRI
Page 191, courtesy of City of Santa Fe
Page 191–192, courtesy of U.S. Geological Survey
Page 192, courtesy of NASA, NGA, U.S. Geological Survey
Page 193, courtesy of U.S. Census
Page 193, courtesy of Environmental Systems and Technologies (A Division of GES)
Page 193, from ESRI Map Book Volume 20, page 47, courtesy of South Georgia Regional Development Center and Valdosta Police Department
Page 193, courtesy of Wunderground, Inc.
Page 194, courtesy of U.S. Geological Survey
Page 196–197, from ESRI Data & Maps 2008, courtesy of ArcWorld Supplement and U.S. Geological Survey
Page 200–201, courtesy of TNRIS
Page 200–201, courtesy of U.S. Census
Page 201, Landsat imagery courtesy of NASA Goddard Space Flight Center and U.S. Geological Survey
Page 201, courtesy of U.S. Geological Survey
Page 201, © 2009 Oregon Metro
Page 202–203, courtesy of TNRIS
Page 206, courtesy of TNRIS

Chapter 8

Page 212–213, courtesy of City of Santa Fe
Page 216–219, courtesy of City of Santa Fe
Page 221–231, courtesy of City of Santa Fe
Page 231, courtesy of U.S. Geological Survey
Page 232–233, Dewberry & Davis LLC, Ken Logsdon, Jr., Technology Solutions Department Manager/Associate

Chapter 9

Page 234–235, courtesy of U.S. Geological Survey
Page 237, from ESRI Map Book, Volume 20, courtesy of Janet Akinyi Ong'injo, Landsat Enhanced Thematic Mapper, Landsat Multispectral Scanner, Tanzania Meteorological Agency, United Nations Environment Programme, and U.S. Geological Survey
Page 237, from ESRI Map Book, Volume 23, courtesy of Paul Schneider
Page 237, from ESRI Map Book Volume 24, courtesy of the U.S. Census Bureau, Population Division
Page 239–240, courtesy of U.S. Geological Survey
Page 241–242, courtesy of NOAA/NWS
Page 245, U.S. Geological Survey, NOAA/NWS, NOAA/NCDC

Chapter 10

Page 246–247, courtesy of City of Santa Fe
Page 249–257, courtesy of City of Santa Fe
Page 262, from ESRI Data & Maps 2008, courtesy of ArcUSA, US Census, and ESRI
Page 264–265, Courtesy of Public Service of New Mexico

Chapter 11

Page 266–267, courtesy of U.S. Geological Survey
Page 268, courtesy of City of Santa Fe
Page 269–270, courtesy of U.S. Geological Survey and City of Santa Fe
Page 271, courtesy of City of Santa Fe
Page 273, courtesy of Tele Atlas North America
Page 275, courtesy of U.S. Geological Survey
Page 278, courtesy of City of Santa Fe

Index

workspace
  refers to a geodatabase 279
World Geodetic System of 1984
      (WGS 1984 or WGS84) 22

X

XML
  sharing geodatabases 11
x, y coordinates
  georeferencing with coordinate
      systems 28
XY resolution 30
XY tolerance 29

Z

zenithal projections. *See also* planar
      projections
zonal functions
  of rasters 205
zoning changes
  application of geocoding 181
Z resolution 30
Z tolerance 29
z-values
  for line features 56
  three-dimensions 63

Related titles from ESRI Press

Getting to Know ArcGIS ModelBuilder

ISBN: 978-1-58948-255-5

With recent advancements in ArcGIS desktop, ModelBuilder has become a visual programming environment that can be integrated with scripts to accomplish complex tasks. *Getting to Know ArcGIS ModelBuilder* introduces users to the interface by presenting basic concepts, and demonstrating best practices through hands-on exercises. This book was written for GIS users who want help automating tasks or performing complex analysis for a more efficient workflow.

Designing Geodatabases: Case Studies in GIS Data Modeling

ISBN: 978-1-58948-021-6

Designing Geodatabases outlines five steps for taking a data model through its conceptual, logical, and physical phases—modeling the user's view, defining objects and relationships, selecting geographic representations, matching geodatabase elements, and organizing the geodatabase structure. Several design models for a variety of applications are considered, including addresses and locations, census units and boundaries, stream and river networks, and topography and the basemap.

GIS Tutorial 3: Advanced Workbook

ISBN: 978-1-58948-207-4

GIS Tutorial 3 features exercises that demonstrate the advanced functionality of the ArcEditor and ArcInfo licenses of ArcGIS Desktop. This workbook is divided into four sections: geodatabase framework design, data creation and management, workflow optimization, and labeling and symbolizing. *GIS Tutorial 3* was designed to be used for advanced coursework or individual study.

Lining Up Data in ArcGIS: A Guide to Map Projections

ISBN: 978-1-58948-249-4

Lining Up Data in ArcGIS: A Guide to Map Projections is an easy-to-navigate, troubleshooting reference for any GIS user with the common problem of data misalignment. Complete with full-color maps and diagrams, this book presents techniques to identify data projections and create custom projections to align data. Formatted for practical use, each chapter can stand alone to address specific issues related to working with coordinate systems.

ESRI Press publishes books about the science, application, and technology of GIS. Ask for these titles at your local bookstore or order by calling 1-800-447-9778. You can also read book descriptions, read reviews, and shop online at www.esri.com/esripress. Outside the United States, contact your local ESRI distributor.